UKRANIAN STOLEN YOUTH OF WAR

OLGA NUNN

Printed in the United States of America.

Library of Congress Control Number: 2022948314

ISBN	Paperback	978-1-68536-905-7
	eBook	978-1-68536-906-4

Westwood Books Publishing LLC
Atlanta Financial Center
3343 Peachtree Rd NE Ste 145-725
Atlanta, GA 30326

www.westwoodbookspublishing.com

Love

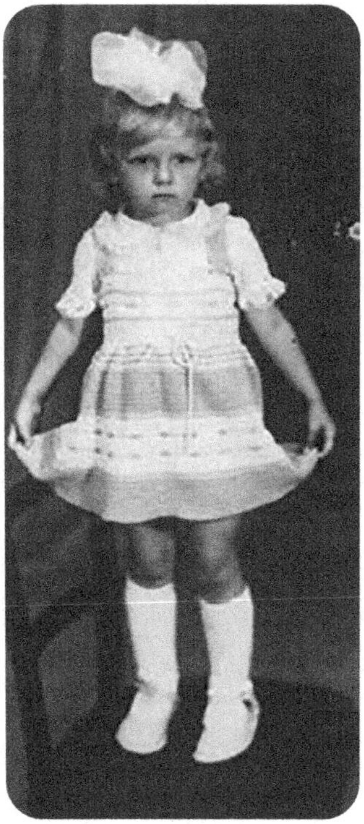

PART 1

A Small Charming Village

CHAPTER 1

This summer was different. I decided to visit the country where I was born. We bought tickets and flew to Ukraine soon after the children had finished their school year. It was so important to me to keep up with relatives living abroad, most importantly my nan, who was eighty-three years old. Every time I saw her, I experienced an amazing burst of happiness.

As I walked towards her where she was sitting in the garden, she tried her hardest to get up from the wooden bench made by my granddad all those years ago. It was clear that she wanted to greet us all. It was so lovely for my dear nan to see her greatgrandchildren. She was always happy to see them; her face just lit up with a wide smile.

This summer was terrific. The weather was warm, and the days were full of exciting things to do. It was a trip that I had been looking forward to for a long time. But work commitments and family obligations had stopped me from visiting my relatives as often as I wanted. The family business and my children's schooling were on the top of our family's agenda. This time, however, I put all my commitments on hold and decided to take my family to Ukraine for the whole summer. Now I was sitting next to my nan, holding her fragile hand, and having a conversation with her. We have always been very close. Even after Granddad passed away and I got married and moved to another country, I never stopped being in contact. Distance didn't matter, and it was always exciting to visit as often as I could.

Later that evening, as we sat in the garden and watched the children playing, my nan started telling me the story of her youth.

This was our special time together, and I felt as if I was back in time with her. When I was a child, she and Granddad used to tell me a lot of stories. It was always very interesting to hear about the war and how they met, including the romance, love, separation, and their close, true feelings. It was a very captivating story.

Nan was quite old now, and from time to time she would stop to catch her breath or to think back to some events so she could remember as many details as possible. Nan's facial expressions changed as she spoke, and as I watched her, the signs of the feelings she was experiencing were clear. Every line and crease in her face looked tense as she spoke. I didn't interrupt when she was speaking; I simply tried to write it all down. I believed this story should be shared with the world. I could vividly see the descriptions of places, people, and events. Soon I was sketching her story in my notebook, and it was just flowing. She began describing the village where she had been born and spent her childhood. The story expanded as I wrote everything down.

In a small village called Babrichi in Ukraine, with only 127 houses, life was in full blossom. The village's beautiful green fields, forests, singing birds, and fresh air gave the villagers a sense of freedom, peace, and tranquillity. The surrounding beauty of nature made it very difficult to take one's eyes off its charming magnificence.

Village streams and lakes were filled with water so clean and pure that one could see one's own clear reflection. The lakes were filled with fish, and the local children loved to sit on the riverbank with their rods ready to catch their supper. This natural splendour cannot be described in just a few words. One needed to be there to appreciate and to discover the true beauty of this small place.

The people of Babrichi started their daily routine from the moment the cockerel sang his first song, and their routine finished when the sun went down. Most villagers were farmers who did a variety of jobs on the state-owned farms run by the Soviet government under

Stalin's leadership. Their daily lives revolved around working for the state, growing their own vegetables, and feeding their own livestock to bring food to the table. This lifestyle had passed from one generation to another and had become the everyday life of Babrichi's people.

The citizens of Babrichi helped one another with the little they had. For any occasion, whether it was a birthday, a wedding, or a funeral, they were there for each other. Not only did the Babrichis work together, they celebrated and grieved together too.

On a bright and sunny day, a young girl called Lina was working in her home garden, picking the weeds from the rows of vegetables. This was my nan. It was a hot day for this job, but compassion-driven Lina was trying to help her parents. She was of medium build and had dark hair that was nicely tidied away at the back of her neck. Her green, beautiful eyes, white complexion, and pink, smooth lips would draw attention from anyone. The simplicity of her outfit, along with her natural features, underlined purity, youth, gentleness, and femininity.

Her parents were working long hours on the state farm. This work was compulsory, and there were heavy state penalties if anyone refused to work. This time of the year was always busy with lots of jobs needing to be done. Lina was the one in charge when her parents were at work. She looked after two brothers, one of whom was just three months old. She had twin sisters who were eleven. Her brother Ivan was seven.

She was working fast today because she wanted to get as much done as she possibly could. She decided to get more rows done and to complete her job more quickly than usual. She wanted at least forty rows to be cleared of weeds, and she was planning to do this as soon as her baby brother was asleep. He had been crying a lot that morning, which was unusual. This was a cry that had echoed right through her. To calm him down, she had given him a dummy—a pacifier—made from breadcrumbs wrapped in a piece of thin fabric. He had sucked on it, and after five minutes, he had slowly drifted off to sleep.

Lina turned around to see where Ivan and the twins were. She spotted their blond heads behind the bushy hedge at the front of the house. "Olya, Nina, and Ivan, come over here," she commanded. "I really need your help today. We need to have these rows finished before noon, as Mum will come home soon and we must show her our good work."

"Just show us what we must do," Ivan replied, and the others agreed with enthusiasm.

Grabbing firmly onto the weeds, Lina yanked them out, leaving the rows nice and clean. The others joined her straightaway, all enthusiastically listening to the instructions from their older sister. Before lunch, the planned workload was completed. The rows were clean, as if the weeds had never been there in the first place.

Lifting their heads up, they all noticed their mum walking towards the house. It was her lunch break. Yarina was a very slim woman of medium height. Her posture wasn't straight; all the years of hard physical work had taken their toll. She was wearing a blouse, a skirt, and an apron. Her hair was covered by a flower-patterned scarf.

"Go, you two, and bring Mum over here so she can see our work," Lina said to the twins.

They quickly ran towards their mum, hugging her thin, fragile body when they reached her. "Mum, we want to show you what we've done today!" Olya shouted.

"Oh, dear children!" Yarina was smiling. She lifted her hands as she admired and appreciated the work of her children and proudly demonstrated her affection for them. But suddenly the moment was interrupted by the scream of the baby. He was wide awake and due to be fed. "I love you so much, my sweethearts," Yarina said. "I really don't know what I would do without your help." She hugged her children with much love and adoration before hurrying into the house to feed the little one.

"Okay, all right, I'm here now," Yarina pronounced as she picked up the baby and put him to her breast. He went quiet again. Lina's mum was feeding the baby on the go while working in the field. Sometimes Lina would take the baby to her mother to be fed and then take him back home. It was not easy for her mum, and Lina was doing everything she could to help. Words couldn't describe how difficult it was for Yarina to leave the baby, but she could not take any time off from work. Every woman in the village was in the same position. Older children had to look after younger children. All women had to work, often giving birth in the field. The newborn baby would be wrapped in a blanket and put nearby while the new mother would carry on with the work. These women were incredibly strong and miraculously healthy. Nothing could break Soviet women.

The clouds gathered together in the sky above and merged into a thin veil that slowly thickened. Lina could hear a distant rumble. Thunder was on its way. In a few moments, a violent clap of thunder shook the landscape. Soon the first raindrops started to fall, and within a minute, torrential rain was pouring on her. Lina had just finishing weeding the last row, and she started running towards the house. By the time she got inside the house, she was soaked. After drying herself and changing into clean clothes, it was time to set the table for lunch.

"Is Dad coming for lunch?" Lina asked her mum.

"He is," she replied from another room while still feeding the baby. "Saw him at work today. He said that he had an hour or so at lunchtime, so put out a plate for him too."

"He must be hungry working all those hours."

"Go to the cellar and bring the cherry juice out."

As Lina was coming out from the cellar, she saw her father walking towards the house. Her father was a man of medium build. He had very strong hands and a good heart. He worked all hours of the day to provide for his family. Maxim was well respected by the villagers. Lina loved her father with all her heart. "Dad, Dad!" she shouted as she hurried towards him.

Her father's face lit up when he saw his little helper. "Now what did you cook?" he asked, waiting for the answer.

"Go inside the house. It's ready, Dad."

"Well cooked, daughter." Maxim proudly praised his daughter. The lunch was very modest. Maxim took a piece of bread. It was dark and surprisingly moist. This would be the first proper meal he'd had for a month. He felt his stomach being warmed by the food.

Lina had learnt to cook at a very young age. She knew how to prepare most meals, and today she had cooked potatoes with herbs. She had made a salad and fried fresh eggs.

When lunch was over, Lina's parents left for work. Lina was tidying up the table while her siblings were taking care of the dirty dishes, washing and putting everything neatly away. Once it was done, Lina took her siblings out to get some fresh water from the well. Singing and laughing, the barefooted children ran around trying to catch one another as they hid behind the trees and hedge alongside the path.

The day flew by, and in the evening, lying in her bed, Lina was thinking about the whole day. She had a sense of pride for being in charge. She felt valued and greatly appreciated in the circle of her family. Her sense of responsibility came to her naturally. She sat in her bed looking out of the window. Before her eyes was endless night, and each star shone its own light. These stars drew her attention closer making her feel as if she was in some kind of magic kingdom. As she observed this magnificently eye-catching beauty, Lina's eyelids became heavier and heavier, and she slowly drifted off to sleep, leaving behind the boundlessly magnificent cosmic view.

In the early hours in the morning, Lina opened her eyes. She could see the rays of sun coming through the window, and the sound of the birds singing made her want get up and get outside the house. Looking out of the window, she couldn't believe how beautiful, bright, and spectacular it was outside. The warm sunlight and the sound of the ticking clock in the background combined as if alerting her for something. After putting her blue-flower-printed favourite dress, brushing her hair, and neatly plaiting it, she was ready to go. She could hear her parents talking quietly. They were already sitting at the table eating their breakfast.

"Come here. Eat with us!" Maxim said, making a room on the bench. "I have been talking to your mum, and she has told me what a great helper you are." Maxim gently tapped his lovely daughter on the back. "We decided that, for all this hard work, we should reward you. You asked us before about this. You can go this Sunday to the local dance."

Lina opened her eyes surprised. The news was so unexpected she could not hide her astonishment and excitement. She had been waiting for months to do this. The day had finally come. *How great*, she thought. "Thank you!" she replied with a huge smile, showing her perfectly arranged white teeth. She hugged each of her parents with much love and appreciation.

"Dancing is this Sunday. I saw the note on the door," Maxim said as he got up and was ready to leave for work. "Lots of work right now." He walked to the cradle. Lifting the little boy, he kissed him gently on the head. Then he carefully put the baby in the cradle and walked into another room to kiss his sleeping children before leaving the house. Shortly thereafter, Yarina left the house too.

Once again Lina was in charge of the entire household. Full of enthusiasm, she started her work early, imagining how wonderful it was going to be to go dancing this Sunday. She was so excited and full of energy to do her tasks for the day. The baby started to cry so she picked him up and rocked him until he calmed down and fell asleep. She slowly put the baby back into his cradle. The siblings were awake now, and after giving them some breakfast, Lina made sure that everyone had a job to do. Everyone had to contribute for his or her share of food. This was the discipline in the Uschuk family. The family lived modestly, but there was always food on the table. Maxim always made sure that his children never ever went hungry.

The Ushchuk family was very much respected in the community for giving a helping hand. Yarina would often take milk round or give some grain to those with less. Maxim would go and give a hand with repairs. Lina and the younger siblings would often take milk round. Helping others in the village was a way of life in Babrichi.

Today Lina was busy with her little brother. The baby boy cried all morning long, and it was worrying. She tried everything—handmade dummy, rocking him in the cradle—but nothing seemed to work. She was seriously concerned because the baby continued to cry despite everything she did. Usually he would stop after she gave him a dummy, but today nothing would make him stop crying. She couldn't wait until her mum was home for lunch. The baby was due to be fed.

"He's crying so much today," Lina said with concern as she met her mother in the yard. Lina held her brother in her arms.

"He might just need to be fed again." Yarina gently took the baby from Lina and went into the house. Her body looked so thin and worn out by the work as well as from nursing her baby son. She looked exhausted from the busy life that kept her always on the go. Her husband was the same. Lina saw her mum being always busy always working, doing something. She loved her dearly and always wanted to help to make life somehow easier.

Today Lina couldn't help but feeling anxious about the little brother. He was doing poorly, and something definitely was bothering him. Lina wanted to ask the village healer if she knew what was wrong with her baby brother. She walked across the street straight into her neighbour's house.

"What brought you here?" the woman asked with curiosity.

"Can you come to our house?" "The little one will not stop crying." Lina was speaking with great concern.

"Oh, I see. I will be round this evening," the old woman replied.

The healer came in the evening just when Yarina was feeding the baby. The healer watched the baby for a while. Lina noticed a look of concern on the neighbour's face. "He doesn't want to take the nipple firmly," Yarina said.

"He is not well," the woman said firmly. "There is a lot of the typhus going around. I have seen other babies who were affected by the virus, and the first sign of this illness is when the baby loses the strength to suck onto the nipple."

Lina and Yarina looked at each other. At once Yarina asked, "What can we do? What can we give him?"

"Oh, just keep feeding him and carry on as you did before." The woman shook her head. "I'm going to bring some herbs to help him sleep. They should calm him down and help him to gain strength." The healer got up and left the house.

The next day, when Yarina's lunch break was over and she needed to return to work, she said to Lina, "Bring him to me if he keeps crying all the time."

The healer returned with the herbs. She explained to Lina how to prepare the remedy. "Don't forget—once the water is boiling, simmer the herbs for no more than five minutes. I'm going to pray for him," she said as she left Lina to care for the baby.

In no time Lina prepared the herbs. After cooling the mixture following the healer's instructions, she gave the medicine to the baby. Afterwards she gently rocked him in her arms. He slowly drifted back to sleep. Lina watched how his little chest was going up and down and his little face was shuddering in his sleep. Lying in his cradle, he looked very peaceful. She covered the cradle with a light fabric to keep the flies off and quietly went out of the house.

A pile of clothes was ready to be washed. She picked up the basket and headed towards the lake. A beautiful landscape of colourful fields was visible from the top of the hill. On the far horizon she could spot silhouettes of the people working in the fields. Turning, she could see Ivan and her sisters playing near the lake, jumping from the stones into the crystal-clear water.

"Ah, you little rascals, this is where you all are? This is why I couldn't see you around the house. Mum was home, and you were nowhere to be seen. Okay, now come over here and help me do this laundry. Quickly!" she ordered. "You can play after it's done. Quickly, over here!"

The children were laughing, but swiftly obeyed their older sister's instructions. Lina washed the clothes by rubbing the fabric against a stone. Ivan was happy to rinse it and put it back into the basket. The whole job was done in no time.

When the laundry was done, Lina rushed back home to see how her little brother was doing. The house seemed really quiet. Lina didn't like the quiet, so she quickly walked towards the cradle and looked at the baby. Her heart was beating hard. She looked through the thin fabric that covered the cradle. He was still sleeping silently with his arms wide open. Not wishing to wake him, she tiptoed outside.

As the weekend drew closer, Lina was looking forward to the dancing event. She had waited for ages for her parents' approval, and now she would be allowed to go. Lina was so very excited, and she was counting days. The dancing normally took place in the local church, and all the youngsters went there on the weekends to enjoy the social event.

The night before the event, to make sure that her beautiful long hair would be shining, she washed it in herb-infused water so that it would smell of the freshness of the meadow. She curled it by rolling it up with newspaper and securing it string. The herbs would make her hair silky, fresh, and shiny for the following evening. She had an early night and couldn't wait for the day to come.

In the morning, Lina woke up in good spirits and started helping her mum in the kitchen preparing breakfast for the rest of the family. The whole family was troubled by the state of the little one. He had a fever, and a rash had appeared on his skin over the past few days.

Yarina kept giving him the herbal tea and was trying to feed him, but the baby was more reluctant to be fed with each passing day. He ate very little. The herbal remedy kept him calm, but it seemed that pain was still bothering him.

Lina knew that the baby had a very serious illness. Several people, including children, had already lost their lives in the village. Their only hope was that the baby would fight through his illness and beat the deadly virus. Everyone was praying.

Lina looked at him, thinking how innocent he looked wrapped in his blanket and how sad it was for him to be unwell. Most of the time the baby cried from the gripping pain. She would rock the cradle slowly until he was asleep again, tired from crying.

The day she had been waiting for had finally come. She stayed busy as if to make the day pass more quickly. Thinking about the evening, she could not wait until her friend came round and they both walked to the local church for dancing all evening. This was the first time she had been allowed to go out and spend time with her friends. Now she had turned into a fine-looking young lady. Tonight she would be able to forget about the work and dance the night away. "The last job is done!" she said loudly.

She started to get ready. The time was drawing closer to six o'clock, and her friend was going to come round at seven. Busy and in a hurry, she prepared her shoes, dress, and her hair. Finally dressed, she looked at herself in the mirror and smiled to herself. Her hair was beautifully arranged and loosely spread over the shoulders. The dress underlined her beautiful figure—a thin waist and curvy hips. Her shoes shone from all the polishing. She looked dazzling and amazingly elegant. Her heart was racing with the excitement. Tonight was a very special night for Lina, and she was ready to enjoy it. She was ready to leave the house, and she wanted to show her parents how beautiful she looked.

"Come here, Yarina," Maxim called his wife. "Not long ago I remember you running all over the meadow chasing the butterflies. But look at you now. You are turning into a very fine young lady," Maxim said with a smile.

Lina smiled back shiningly at her father.

Her mum was quiet. Lina spotted her mum's pride in her tearful eyes. She walked across the room to hug her mother with much love and affection.

Lina's friend was already at the door. She was also dressed up in her best clothes, and her hair was beautifully styled. Both girls looked amazing. It was already getting dark. "Let's hurry!" They started to laugh.

Most of the village people were in their houses sitting around their tables. The girls peeked through the windows of the houses on their way. In no time they were near the church. It was the place where people gathered. Someone was already playing an accordion, and they could hear the sounds of everyone inside. The girls walked in, and everybody looked at them. Lina was blushing.

"Hi, girls, how long have you been here?" Lina asked approaching a group of her friends.

"Not long," answered one of them. "We just got in before you. Can you see that Stepa is playing an accordion tonight?"

"Good," said Lina's companion. "I'm glad it's him. The other boy from the nearby village was useless. What's his name?"

"I don't even remember his name," said another girl. "Anyway, Stepa is the best musician by far."

"Tonight is our lucky night, girls," Lina exclaimed, and everyone laughed.

"The music is going to be great!" said her friend. The girls hugged each other. Happiness shone on all their smiling faces. They were full of joy and excitement. This was a truly magnificent start to the night. The room would be filled with good music, they would be surrounded by friends, and there would be dancing and lots of fun. In the dimmed lighting, the moving silhouettes of the youths on the dance floor were reflected by the dancing shadows on the walls of the building.

Lina looked across the dance floor and saw a group of friends who were dancing. "What we are standing here for?" Lina said with much enthusiasm. "Let's go!" She pulled her friends onto the dance floor. The dynamic melody of the accordion made her feet wanting to move to the rhythm of the music. She just couldn't stand still. Her feet began to move to the music, and the rest of her body enthusiastically expressed the dynamic rhythm of the music. Every single emotion showed in the way that she moved her body. She lost her self in the cheerful melody of the accordion.

No one could dance like Lina. She threw herself into the music, looking absolutely beautiful. Her dazzling movements made her shine in the dim lights of the church. Everyone was looking at her, but she could not see it. Her feet were moving in the accurate but rather complex pattern that only she was able to dance so naturally. She had been born with it, and no one could dance the way she could. The rhythm of the music went hand in hand with the way her whole body responded. She looked amazingly beautiful and incredibly graceful.

The young people danced the night away until the early morning rooster crowed. Everyone began leaving the place. They sang all the way home with hearts full of joy and laughter. This was going to be a night they would remember for a long time to come.

But the events that were about to happen would change their lives dramatically to an extent that they would never have dreamed of.

Lina walked into the house quietly so she would not wake up anyone. As she was about to walk into the dining room, she noticed her dad sitting with his hands over his face. Lina's instincts told her that something bad had happened, and a sense of panic grew inside. "Dad, why are you sitting here? What's wrong?" she asked. There was no reply. She walked towards her dad and took his hands away from his face. She saw the pain he was in. "What happened?" she asked in a loud and panicky tone of voice.

"It's the baby," he replied quietly. "The little boy is not breathing properly." He started sobbing.

Lina ran screaming to the cradle. She couldn't believe this shocking news. The pain in her heart was so bad that she couldn't handle it any longer. All the family members were on their feet now. They were stunned by the tragedy of losing their baby son and brother. The ordeal of loss was unbearable, and a heavy shadow of grief came over the Ushchuk household.

Yarina was the strongest one. Her endless prayers had given her a sense of calm, but deep down, she knew that the baby was losing his grip on life. In the last few days, the baby hadn't eaten well, and he had cried all the time. Yarina had not wanted to worry the family members, so she had just kept it to herself as she prayed. It was hard for her to watch her own flesh and blood lose the spark of life. Every day she prayed that her little boy would be saved, but the disease of the growing epidemic had completely taken hold of the little fragile body that still seized every opportunity to live.

"Heaven needed another angel!" Yarina spoke after some time. Her face was swollen from all the crying and pain. Every single night she had cried, hiding her tears away. Her heart ached as she watched her tiny son suffering. His little fragile body couldn't take it anymore. He had come to the end of the little life he'd had. "He won't be in pain anymore." The tears streamed down Lina's face as she looked at the little lifeless body lying so innocently in the cradle. "Now I have to organize the funeral."

"How we are going to be without him?" Lina asked with confusion. She was still finding it hard to come to terms with the fact that her little brother was gone, and she would have to say her last goodbye.

"You have to go around to the neighbours and tell them about the funeral," said Yarina. "I will get everything else ready."

Lina's sisters and her brother were sitting on one of the beds cuddling and comforting each other, their eyes red from tears.

Maxim was speaking now. Everyone in the house turned their faces towards him and the pain reflected all over his face, every little muscle, as he was speaking. "I never knew I would be putting my little fella into the grave," he said. He was sobbing; his entire body shook violently. He could not come to terms with this terrible loss.

Ivan took Maxim's hand. "Let's go outside, Papa." Ivan pulled Maxim by the hand.

The days after the funeral passed by quickly. Lina was occupied with her daily duties. She did all the usual work as if she was trying to focus her mind on something other than the family's loss. She couldn't stop noticing that one of her twin sisters was not well. She was terrified by the disease that had taken the life of her little baby brother.

Her sister's health was deteriorating with each passing day, and the whole family sat beside her bed each evening, praying for her to get better. In the night, they lost another member of their family. Shortly afterwards, the second twin sister contracted the virus. She lived only for two weeks. It was the third horrific ordeal that Ushchuk family had been forced to go through. This terrible virus had taken three young family members—three funerals in two months. The whole family was grieving.

My nan was getting tired and stopped telling the story. "I will continue tomorrow. I am a little tired now." I got up and gently kissed her on the forehead and asked her if she needed anything.

She looked at me, and after several silent moments she replied, "A glass of lemonade, my child. She was diabetic but occasionally fancied a sweet drink.

I decided to leave my writing until another day when she felt like continue with her story. Shortly after her evening wash, she was in bed. I didn't mind helping her with everything because she had done the same for me so many times.

"Good night, Nan," I said, gently kissing her once more. I realized how much this dear person meant to me. She was so special in every single way, and I loved her so very much. That love was stored in my heart and would last for eternity.

Over the next few days, my nan carried on with the story. The plot unfolded further and further, unveiling more and more events. I didn't want to rush her. She would tell me bits of the story, and then she would stop for the moment trying to think before she continued. The events she described revealed so much human tragedy, loyalty, patriotism, and unspeakable bravery. I was astonished to hear about the crucial moments of history as they were revealed to me by the person who was so dear to me. This was the true story of my grandparents, and it explained so much. These moments would stay with me forever. I could vividly see the scenes as I jotted down everything in my notebook.

In the morning, the tradesmen were waiting for Maxim in the local farms office. After giving everyone a task for the day, Maxim mounted his horse and galloped off to take a look at the field where Yarina was working.

"What do you think, Maxim, the war is coming?" One woman asked him as other women gathered around.

"Maybe," said Maxim, "but for now we have to work for peace."

They talked of the war, sensing some tension in the air, the premonition brought by the current talks and the news of an immense German army that seemed to continue to expand and threatened to step on the soil of their motherland any day now.

Right now, Maxim felt good from looking at the blue clear sky that was very high above his head and the sun that was shining onto the crops. He loved this peaceful silence in the village. However, he couldn't help the feeling in his gut that soon this peaceful silence would come to an end. It would be interrupted by loud thunder that would turn everything he observed today upside down. In fact, he didn't feel—it he knew it. He almost knew his fate, as if inside his body right under his heart was already sitting a piece of the bullet that would make him close his eyes forever. And in just two short weeks, Maxim's premonition became a hellish reality.

Ancient natural forces chased the waters of the river to the east where it looped between the grey willows and the trees with their long-lasting crimson leaves. It was as if the river was connected by an umbilical cord to all the wooded areas that surrounded the hamlets where people dwelt. As it had done for hundreds of years, the village of Babrichi was meeting a peaceful sunrise. People were taking their cattle onto the meadow. Men were smoking paper-wrapped tobacco. Women were preparing in the field to bind the sheaves of wheat.

Suddenly the peaceful silence was broken by the crack of flying motorcycles and, after their arrival, the explosion of unusual foreign speech. A group of SS intelligence officers burst into the village. After the roar of the motorcycles stopped, the newcomers began to confer among themselves. Not long after this, the armed group jumped back onto the saddles of their machines. Frightening a flock of crows, they took off from the place. Only one motorcyclist from the group left in the opposite direction.

The unexpected appearance of these uniformed and armed people and their motorcycles roused the people of the village of Babrichi into loud conversation.

In June 1941, the Germans launched their invasion of the Soviet Union: Operation Barbarossa. A convoy of SS vehicles arrived. On the car bonnets, flags emblazoned with swastikas snapped in the wind. This sent a sense of curiosity among the villagers. They were not frightened, but their interest increased. They all stopped what they were doing at the time and came outside to observe the newcomers. The men who got out of the vehicles wore green uniforms decorated with swastikas. The Prussian military obsession had long been overlaid with Nazi formality, including saluting, heel clicking, and the inevitable 'Heil Hitler!'

Usually life in the village was quite monotonous, so the villagers' attention was shifted to the strange appearance of the men in uniforms. Today was the day that would create dramatic changes in the lives of every villager and turn their world completely upside down. They would never have been able to prepare themselves for what would happen.

Until that time, for many centuries, the uneven village roads had served well for horses and carts, but now they were overtaken by the deafening groans and roars of tanks and motorcycles. The fresh smells from the meadow were replaced by the smell of burning petrol. The village's dogs, usually lazy and quiet, sensed the strangers. The dogs were agitated by these unwelcome visitors. Even the oldest and quietest of dogs expressed their annoyance.

Narrow streets that had never been full of people were now too crowded to move through. The tidy fences made from tree branches were cracking from interference from the wheels and sides of the vehicles passing by. Almost at once, as newcomers hauled buckets up from the depths of wells, the chains holding the buckets screeched as if they were unwilling to give away water to the strange invaders.

At first, there was no shooting or bombing. Only the village's dogs, as if possessed, barked madly. They sensed better than their owners the trouble that had come into the little village. Everyone heard the thuds and crashes as the SS men kicked down doors and shouted orders. And if anyone objected—and a few did—the answer was a bullet in the head. The village was full of the sounds of ghetto dogs barking, boots clashing, trucks and motorcycles dashing back and forth. From a distance one could hear the short, sharp screams punctuated by the dull cracks of pistols or rifles. And then it would be quiet again.

"Ba-bri-chi," the German interpreter sounded out the name of the village to the officer in charge.

"Ya, ya," the officer responded, nervously rubbing his chin, trying to repeat the word. They were wearing Wehrmacht uniforms with black epaulettes and lightning flash symbols on their collars. They wore eagle badges on their sleeves. These were the execution squads, tasked by the Nazis to perform systematic attacks on the people.

One could spot on the faces of the invaders their disappointment and annoyance at the village barking dogs. One could see from the villagers' cold gazes, awkward smiles, and grins with hidden meaning that they were not pleased to see the invaders. This annoyed, offended, and dissatisfied the newcomers. This made them angry; they wanted to respond with an evil retaliation.

The deafening barking of the dogs turned into deep growls when the people in uniform approached them. The dogs carried on barking, and terrified children could not be quieted despite the best efforts of their parents. An SS officer, a cruel Nazi, playing judge, jury and executioner, pointed fingers, vigorously gesticulated. A random command was given. It came from the sick mind of a madman.

The *starosta* —head man of the village—who had voluntarily offered his assistance to the invaders, was ordered to catch all the dogs and take them to a registration point. His surname was Bezruchko. He was a person of slimy character who always bossed women and men working in the fields. He thought he was in charge and could do what he wanted. For this reason, the people in the village disliked him. He didn't like anyone other than himself.

Now he was quiet. Wonder why? As he tried to carry out this order, the people were unhappy, and the dogs turned angry. Everything turned into a state of confusion. On the outside, members of the SS only looked polite, but it was difficult to find out what was on their minds.

The villagers couldn't understand. What sort of registration was it? They had never heard in their lives that house dogs should be registered.

In addition, the newcomers noted down all the apple and pear trees, counted all the cattle, and took away all the horses. What the heck were they doing? In the First World War, the cattle had been taken, but now the dogs were taken as well.

"So what about those stray dogs without a home?" Yarina wondered. "Do we have to answer for those too? Do they have to be registered too? Dear Lord, what's happening? I was waiting for St Yuriy's day, but instead the trouble has come here."

"Our Black will have to go?" asked Lina, looking at the sleeping dog.

"Seems that way," Yarina replied.

"Open the gate!" the SS men ordered bluntly.

A person without freedom is worse off than an animal. At least animals can show their annoyance. A cow can buck and jump, putting her tail up showing her dirty backside. But how can two females protest in front of SS men who carry machine guns and hold their fingers on the triggers. And beside them, a huge aggressive German shepherd, who was ready to tear anyone apart, showed his anger right in their faces. A single command would set the dogs off. The females' fate could be decided right there, right at that moment.

The SS officers' dog, who was inseparable from the soldiers, was hated by the villagers and the local dogs. At the first opportunity, the local dogs would rip her apart. She looked heavily pregnant and could hardly walk. The shepherd felt heaviness in her tummy and full hanging nipples. Dribbles of saliva ran from her mouth from the anger, and she hated anyone who came close to her. She was annoyed at the smell of gases and at the movements and thoughts of any strangers standing in front of her. The unfamiliar situation, along with the pointless barking of the other dogs, drove the bitch crazy. She uttered a deep, nauseating growl. The owner became worried about his dedicated companion and the future brood that would be born any time now. The bitch needed quiet, peaceful surroundings with no one disturbing her. And here she was surrounded by these disgusting mongrels. Nothing would interfere with the birth of the real and healthy Aryan breed! So the others must shut their filthy traps permanently!

Early in the morning the show started. The SS were patrolling the streets, armed and watchful. The odd dog strained at its leash and barked at anyone passing by. Some villagers decided to unhook their collars and set them free, pushing and shouting to make them run away, but the dogs' loyalty and dedication took the unkind treatment from their owners as a strange game. They kept running back wagging their tails and playfully rubbing against their owners' legs. But their happy barking was not appropriate at that time and place.

The starosta's dog was furry and bulky, and looked just like his owner. He refused to come out from his wooden house. Bezruchko wrapped the chain around his hand and pulled the dog with force out to the place of registration, setting example to everyone for what to do. No one had a choice but to do the same, thinking that nothing bad would happen. But it did happen!

The people of the village understood what registration was about as soon as they saw what the invaders had organized near the wild pear tree that supported an old ladder from the local farm office. This was where all the important community meetings were held, and this was where it had been ordered that all the villagers should bring their dogs.

The barking, screaming, lively dogs, who were used to seeing crowds of people in the local farms office, could not understand why everyone was there so early in the morning and why these people in uniforms equipped with guns had locked everyone in a circle. Where had all the swings gone? Why was there a deadly silence? And what was meant by those loops made from ropes, hanging with laziness from the branches of the trees.

While the officer in charge was giving everyone orders about what to do with the animals, the interpreter translated. Everything happened as if it was a strange film. When the dog Bala was ordered by the officer in charge to be put into the rope loop, a ginger major sergeant took the dog from the shocked owner. Bala twitched with convulsions and soon after stretched in silence. The cold hands of deadly fear grabbed the throats and thoughts of the villagers at once. Hands and legs refused to move, and the people were unable to understand what was happening.

Then all the other dogs, one by one, were hanged, disturbing members of every household. The dogs had been loyal guards, shepherds, hunters, and pathfinders. They had remained strong through hunger and cold, whether treated badly or kindly, whether beaten by abusive owners or treated kindly by others, whether trained or untrained. Bobik, Tyzik, Pirat—all of them were destroyed without consideration for their colour, breed, or family tree. The wild pear tree rocked from the heavy weight on its springy branches.

Some dogs had time to answer with the snarl. Some licked the hands of their owners, and some tried to catch with their teeth the sleeves of their executors.

The tree had turned into hanging point. There was squeaking from the weight on the branches as its terrifying harvest ripened in front of the owners' eyes. They looked like screaming bells. In front of the speechless terrified villagers, the dogs died without losing their loyalty until their last moments. This was dogs' devotion, and they didn't have enough time to understand or to forgive their owners who had been ordered with force to stand back without making any movements whatsoever.

The last one on the list was the brigadier's hunting dog, whose fur was a mass of colourful patterns. Suddenly, to everyone's astonishment, the brigadier instructed his dog to attack. The dog started to run, and he jumped onto one of the officers who had been performing the executions. This act decided the brigadier's fate. He had been foolish, and he paid for it with his own life. Everyone heard the burst of machinegun fire, and the senseless spectacle came to an end. The people gasped with fear and ran away in panic. The dogs that had not yet been executed by hanging were shot on the spot. The SS invaders didn't touch the villagers. The order had been carried out according to the plan.

The village of Babrichi seemed as if it was dead. There was no light. There were no barking noises from the usual occupants.

The SS men stood by, petrol cans in their hands, admiring their handiwork. They were about to burn the evidence—the tree and the hanging dogs—in a desperate attempt to conceal their crime.

In the night, wolves surrounded the entire village. Yes, yes, their dreadful howling made the villagers' skin crawl. The howling, at the same time, disregarded every rule that prohibited the disturbing or annoying of the new arrivals. It was June, the beginning of the summer, the most exciting time of the month for the forest predators. Without an explanation, the animals were gathered as if by some invisible force. They reminded everyone of their existence with their dreadful crying and howling, which reminded the villagers of harsh winter storms and freezing winds they had experienced in the past and that were still yet to come. Or could it be that the sudden appearance of the wolves was their way of paying respect to their domestic fellows who had lost their lives so ingloriously? Who knows? Who knows?

Lying in her bed, Lina kept hearing the noises of the wild predators. She thought how strange it was to hear them this time of the year. In the dark winter nights, it was common to hear the wolves' wild howling, but not in the summer months. *Maybe they came to pay their respects to the executed dogs*, she thought to herself. Lina's skin was covered in goose bumps from the dreadful sounds and the horrific scenes still vividly fresh in her mind. It was a worrying night for all occupants of the village of Babrichi. Frightened by having to watch the dogs' executions, as well as by the unprecedented invasion of wolves, Lina tried to get the dreadful thoughts out of her mind.

That night in their houses, the people of Babrichi whispered about the unexplained comets in the sky and other strange and mysterious events that had happened just before enemy invasions in the past. This current sudden and unexpected invasion had brought to the peaceful village of Babrichi complete chaos, which left vivid scenes of horror in the minds of each villager. On this night, the residents of Babrichi were not thinking about their daily routines; rather, they were thinking about the invasion of uninvited strangers who had stepped onto their land and forced atrociously evil rules.

No one knew how far the SS wanted to push and what other mad ideas they had in their heads. No one knew what was to come, but the spirit of the villagers would not be broken. Most villagers had already dealt with devastation and loss of relatives caused by epidemics, hunger, poverty, and now the uninvited invaders. The events of that day made people aware of the coming storms of war.

In one of the houses lived the officer in charge with his loyal and very-much-loved dog friend. Something weird was happening. The bitch was in labour ahead of her time. Shocked by the pain and fear of the depressing howling of the wolves that seemed to get through every crack in the house and chimney, as the pups were born, the young bitch went mad and strangled all of her children. She ate all four of the newborn puppies. Disappointed by what his dog had done, the officer, finding it hard to control his nerves over the embarrassing tragedy, could not handle the situation any other way—he shot the poor creature. *How wild and scary this country has become!* the officer thought in despair. He could not get out of his mind the shine that had appeared in the eyes of the village people after the dogs' execution ended. Their stern looks were on his mind, and no matter how hard he tried not to think about them, their faces continued to stare at him. He could not fall asleep. His peace would be disturbed forever. This was just the beginning of what was to come, and it was only the second day in the village of the most disastrous and eliminative in the human history of war.

My nan stopped to catch her breath again. She had been telling the story for a while, taking short breaks. Some things she didn't remember, but what she did remember she had been telling me in great detail so that I could note everything down and then bring to life the events of the past. It would then be up to me to tell the people of today about the human survival, tragedy, and much more. Nan slowly drifted to sleep. After I put the blanket over her legs, I left her to rest.

PART 2

Lina Is Taken

CHAPTER 2

It was the middle of the night. In the centre of the room was a heavy, rough wooden table. On the table was a simple kerosene lamp that reflected yellowish shadows on the lightly painted walls. The family sat on benches on either side of the table as they shared a meal. The table was tidily covered by a hand-crafted tablecloth. In the centre sat a vase overflowing with fresh blue flowers collected from the meadows.

In each of the corners of the room hung hand-crafted towels and icons of saints. The centre of attention was the main icon, which depicted Mary with the baby Jesus. The cottage looked cosy.

After the meal, everyone went to bed. After they had gone to sleep, only occasional snores broke the silence inside. The crackling hot ashes in the stove kept the children warm in the night. In the next room their parents slept.

Loud footsteps disturbed this peaceful silence. A sudden heavy knock on the door woke Maxim up. In no time he was on his feet, quickly heading to open the front door. "Who is it?" he asked loudly.

"Open the door!" the voice on the other side ordered.

Maxim quickly recognized this voice and knew that it brought bad news. SS men and the starosta walked in and ordered him to wake everybody up. Envy lay behind the silent starosta's gaze, and his sarcastic smile showed his tobacco-stained teeth as he walked into the Uschuk family home. He was a monster. His only achievement was grassing on his own people and stealing and embezzling all they owned. How could anyone think of him as good when he deliberately sent the youth to SS labour camps, supplied the SS with information about the people in Babrichi and the nearest villages, and dishonoured as many as possible, simply for his own advancement. Half asleep, Lina sat on the edge of her bed. She couldn't understand what was going on as she tried to put her clothes on. Her entire body was shivering from this sudden uninvited visit. Now she knew they had come for her! She could hear the sobs of her mother as she was preparing some food and other bits, wrapping it all up in old paper.

"What are you staring at?" the praepostor sharply asked Maxim.

"Give us drinks and some food," the starosta demanded. "And hurry up. You know the officers don't like to wait. Today she will be taken. She will be transported with all the others to Germany."

Maxim turned to Lina, who was standing behind him. Putting his hands on her shoulders, he said in a concerned tone of voice, "Lina, daughter, remember—don't run and don't steal! Remember this, my child."

Lina saw her mum wiping the tears that streamed down her face. Yarina walked across the room gently and gently hugged her daughter. She didn't want to let Lina go, but Yarina's fate was already decided; she had already lost three of her children. Now these beasts had burst into their family home at night and were taking another of her children, one of the two that had survived. Yarina wanted to tear these bastards apart, but she knew they would all be shot on the spot. So, clenching her teeth and fists, she prayed for her child. "I've wrapped some things, my child." Yarina was sobbing.

Lina grabbed her dear mama with both arms and clenched her, not wanting to let her go. "I will be back," she said, sobbing. Lina was hurting and didn't want to go with the strange men in uniforms who had invaded her land and were now standing in her family's home tearing all of them apart. She couldn't refuse either because her family was at risk. Having no choice, she obediently left her home. Her fate was sealed in the hands of the harsh and cold beasts.

The sound of her mama's crying was still echoing in her ears. Before her eyes was the image of Ivan's sweet little face with tears streaming down his cheeks. His only sister was being taken, the one who was support for him and for his parents.

The night was filled with moisture, and the smell from the meadows filled her senses, bringing many good memories. She breathed deeply as if trying to keep hold of these memories because she didn't know when she would return. As she walked slowly to the truck parked on the side of the road, she was crying. The "black crow" truck was already waiting, the eagle and swastika emblazoned on the doors.

She thought that, if she cried hard enough, they would let her go, and she would be able to return to her family. But this hope was soon gone. The black crow was taking her out of the village in the direction of captivity. The happy days were long gone, and life would never be the same again. She experienced the sudden realization of being alone and in pain. It was as if a nasty spider had crawled inside of her, penetrating deeper and deeper into her soul, squeezing everything inside. Only warm tears were running down her cheeks relieving the tight knot inside. She prayed in spirit for hope. The rumbling noise of the truck's engine broke the silent night as it took her further and further away from home and her family.

After a long time, the truck suddenly stopped, and the sound of the doors at the back being unlocked indicated their arrival. Lina jumped onto the ground from the back of the truck following the orders to walk towards a nearby grain barn. Lina remembered so well the fresh smell of the stored grain. When her aunt was working here, Lina and her cousin often played inside the barn, walking with bare feet on top of the stored heaps of grains.

The sweet memories soon vanished at the heavy sound of the opening doors. Suddenly an awful smell hit her nostrils. She walked inside the dark building, which was full of people cramped together all over the floor. She tried to look to see if there was anyone she could recognize. Suddenly she recognized her cousin, Katya. She was sitting several steps away in the corner of the barn. Lina lightly shouted from the excitement, "Katya, hi!"

Katya jumped up onto her feet and started making her way through all the people. "I'm glad we are here together," Lina exclaimed gently hugging her cousin.

"Come and sit here," Katya said. "Don't be afraid. We are here together, and everything will be much better now."

"Here, take some sweets," Lina offered.

"It's my favourite—goose step sweets!" Katya exclaimed. "Now it's not too scary," she said, hugging Lina.

Her cousin gently kissed Lina on both cheeks with much love. At least they were together in that dark, scary, and dumpy place with other young people. Hundreds of lives were in the hands of the cold-hearted guards outside the barn. The young girls and boys didn't know what the future held for them or how life would treat them.

The barn was now just a dingy place full of nauseating smells and the rumbling noises of quiet talking, crying, and moving people. It all made Lina's skin crawl. They were waiting for a train that would take them to an unknown destination where they would live with hard work and graft. Some of the young people had been there for a month or longer without knowing what was going to happen next.

The morning light revealed otherwise unnoticed cracks and holes in the walls of the building. Lina had just awakened, and she was looking around to see if she could recognize anyone else. She noticed a boy who was sitting with his head down. Not knowing why, she got up and walked to him and offered him some baked rolls. The boy slowly got up to his feet, and with a lazy smile and much appreciation, he took the treats. "I'm Kiril," he said quietly.

"My name is Lina, and I was brought here last night. How long you've been here?"

"So many days have passed. May be a month or longer now."

He was an orphan, Lina found out later, and he looked severely under nourished. She shared with him the food she had. The youngsters who sat around her grabbed her food with greed, thanking her. They were glad for the chance to eat. They shared stories about how each got there, pouring their hearts out to each other. This was the only way they could support one another—the best thing they could do in the given circumstances. They remained grouped together as they went through such an unbelievably weird ordeal. They tried to stay positive; it was the only way to survive the insanity.

Lina told the others quietly how the SS had come into her home with the starosta in the middle of the night. She cried as she described how upset her family had been.

"At least you had the chance to say goodbye," Katya said. "I didn't even have a chance to say goodbye to my parents. The SS men took me as I was walking on the road. Black Crow stopped, and several SS men jumped off the back of the truck. They grabbed me and my friends and pushed us into the back of the track. They were saying, 'Schnell, Schnell!' (quickly, quickly). It was so scary, so sudden, and I could not understand what was happening and where they were taking me. They were saying something, but I could not understand what they were saying. I thought they were going to take me out of the village and kill me with the rest. We realized that Black Crow was moving in the direction of the train station." Katya told all of this with a horrified expression on her face. "Then I thought for a moment, *If they wanted to kill us, they would have shot us there and then.* I thought, *I'll be brave no matter what happens.* Then all of us were brought here where we all are now." She ended her story by raising her dark, thin eyebrows and looking around.

"Me too," a skinny girl in the corner said. "I didn't have the chance to say even a word to my parents."

"The same here," another youngster said.

"And me," a boy with big floppy ears said.

Lina looked around and noticed how many had been deprived of saying their goodbyes. Deep down, Lina hated the beasts in the uniforms. They had turned the lives of all the villagers into chaos and destruction. They had caused pain, tears, and separation.

"Don't worry though, because it's a good omen," the girl with freckles all over her face said. "My mum always says that, if you don't get the chance to say goodbye, it means that you will meet again."

"Hope so," Katya replied.

"Don't worry. Of course you are going to see your family." Lina was trying to reassure her cousin. "They will take us to Germany, and after some time, we will return. You will see. We are going to do everything it takes to make sure that we return home."

"Yes," Katya said, nodding.

"Now don't worry, and don't think about the bad things. Pray for our return." Lina reassured her cousin once again.

Both girls promised to support and to look after one another. Lina hugged her cousin because she wanted to make the fear to go away. Lina didn't want Katya to be upset and was comforting her in a caring, motherly way. She wanted her younger cousin to be okay and to come to terms with what had happened. After all, there was not much they could do to get out of this dingy, damp building or to change anything. Besides, it was very risky to do anything to retaliate. There was no point in doing anything but stick together somehow to get through this dreadful ordeal.

"I was sobbing, but they would take no notice," Lina said loudly so the others could hear. "We just have to wait and see where we will be taken and how life is going to treat us."

CHAPTER 3

In the night Lina woke up because someone was calling for help. She and Katya quickly got up and walked towards the person who was calling out. It was a young girl. She was lying down in the corner and was suffering with a high fever. It appeared that she had some kind of infection. Lina got some water and sprinkled it on a piece of fabric. She put it against the girl's forehead to cool her down. The girl was delirious and so confused that she could not even tell her name. Lina checked on the poor girl on and off all night. Katya and some of the other youngsters tried to help her. The girl's health continued to deteriorate, and they had no choice but to let the guards outside know. They were hoping that the girl would be sent home. They wanted to save her from being transported and give her the chance to be free. The SS men came in with a stretcher, lifted the girl up, and took her out of the building.

The next day the barn doors opened. The parents and relatives of the young people inside the barn were gathered outside the doors behind a barbed wire barrier. They were strictly ordered to stay where they were; the boundary was guarded by SS men and their dogs. The visitors began to shout the names of their sons, daughters, brothers, sisters, nieces, nephews. The loved ones were hoping to catch a glimpse of their children or give them some food.

Anyone who attempted to climb over the wire would be shot on the spot. The people could see that someone had been taken out of the barn on a stretcher, but no one could see who it was. They watched the men carry the stretcher to another building. Everyone was in a state of panic. Maxim didn't hesitate. He went into the village to get the starosta, and demand that he speak to the SS. Maxim thought that it could be Lina on that stretcher, for all he knew. He walked briskly towards the starosta's house. Maxim knocked on the door and firmly pulled the handle after hearing the starosta's reply to come in. He bravely walked into the hallway and, without any hesitation, spoke in the firm voice: "I want to ask you something. I have known you for many years, and we grew up in the same village. I know that the war has put us in the situation that we are in now. In the name of God, I'm begging you to talk to the SS and find out who was on that stretcher they took out. Only you can do this."

"What, Maxim? What do you want?" the starosta asked with irritation and annoyance in his voice.

"I was standing outside the barn and saw the SS guards come out with someone on a stretcher. It could be my Lina. Surely it was someone else's child. Please can you go and find out who it was?"

Suddenly a woman barged into the room and pointed at the starosta. "He helped them to take my son from me. He's the only child I gave birth to. One day he will pay for everything he did!" She cursed the starosta.

The starosta left his house, met up with the two SS men, got into the truck, and drove off.

"See what I mean, Maxim?" the woman continued. "He's a rotten traitor full of crap!" She spat on the ground.

Maxim was determined to find out who was on that stretcher. *I'm not afraid to face the bastard again. I will get it out of him,* he thought with anger.

CHAPTER 4

In the evening, when it was a bit quieter in the village, Maxim was ready to pay another visit to the traitor's house. He headed out to see the starosta. He entered the front gate of the bastard's yard and continued walking firmly towards the front door. Maxim knocked on the door and waited for a while. It was quiet, and no one opened the door. He knocked louder. He could hear someone's footsteps inside. The door flew open, and the starosta's wife was standing in front of him. "Oh, it's you, Maxim! Come on in. What's brought you here? He's just having his evening tea. He doesn't like to be interrupted."

"I'm going to wait outside," Maxim said, looking around the dark corridor.

"Okay, dear," she said with a tired look on her face.

Maxim sat on the old bench while he waited. The emotions he was just about able to control caused him to put his hands into tight fists.

The satrosta's wife finally called to Maxim. He walked into the dining room. Straight across the room, at the big wooden table, sat a huge, fat, greasy example of human deceit.

Maxim spoke first. His speech was short, direct, and to the point. "I need to know who was on the stretcher. Everyone wants to know. People are waiting."

"Maxim, why do you care? It wasn't your daughter, if it makes you feel any better."

Maxim was relieved, but at the same time he was determined to know the name. Deceit was written all over the starosta's face. His nose was the biggest feature on his face. It looked like an out-of-shape potato that had turned purpely-blue over the years. His greasy chin was evidence of the rack of ribs he had just eaten. He had left the bones on the table for his wife to clear up.

"Just tell me the name," Maxim insisted, speaking through his teeth.

"The girl was from our village, and she died from the typhus."

"Who?" Maxim shouted.

"Nina Kornienko, daughter of Pavlo Kornienko."

"Oh, Lord!" Maxim proclaimed loudly. He had seen Pavlo earlier in the day standing with all the others outside the barn trying to give some food to his child. He didn't even know that his child was dead. "Where is her body?"

"The SS had strict orders to burn the body to stop the disease from spreading." He bluntly looked at his watch. "The body has already been burnt."

"The family deserves to get the body back so they can say their last goodbyes." Maxim was firm.

"Maxim, what do you want? It's war, man. What do you expect?" The starosta was shouting nervously. Spit was shooting out of his mouth. His nose looked as if it was going to rot soon and fall off.

"How I'm going to tell this to Pavlo?" asked Maxim. "How do you want me to say this to him?" Maxim had tears on his eyes.

"Don't tell Pavlo or anyone else about this conversation, do you hear me?" the starosta replied with great annoyance.

"No, I can't lie. Pavlo must know the truth."

The starosta could see Maxim's anger. Even then the man couldn't do the right thing. He was so far gone from honesty and decency. "Leave my house now, or I'm going to call the SS." Fear was printed all over his face. He was pale and ill looking and didn't want to be alone. He could almost smell the danger so close to him.

"I am going to leave, but remember these words," Maxim said. "You are never safe anymore."

"Don't threaten me!" The starosta was shaking nervously; Maxim could see that he was full of fear and deceit.

Maxim walked out of the house and headed in the direction of Pavlo's home. Pavlo was sitting on a bench when Maxim walked in. "Maxim, what has brought you here?" Pavlo asked with concern as he noticed Maxim's pale face. Pavlo's wife, Darina, came out from another room holding a pile of laundry.

Maxim felt pain over what he was about to do. He had to deliver such terrible news. "I have just come from the starosta's house," he began. Suddenly everything went quiet. Pavlo and Darina looked at each other. They knew it was about their Nina and it was no good. Darina was hysterical. Her hands were shaking violently, and Pavlo was holding her to keep her from falling to the floor. The woman could not control herself. She was sobbing loudly, screaming, and cursing.

Maxim couldn't open his mouth to tell them what he knew. Pavlo ran into another room. Darina was on the floor, hitting the floorboards with her fists. Pavlo appeared with a gun on his shoulder. He ran out of the house. He jumped on his horse and galloped into the night. Maxim didn't have the time to stop him. It was too late.

"Come on, Darina. You come and stay with us, dear." He stretched his hands to comfort.

"I don't want to live right now. How could they do it to her?" Darina was sobbing violently.

"Let's go. Yarina will talk to you. Come on, love. Come with us." Yarina would know what to do. He helped her to climb onto the cart, and then he sat at the front. He took up the reins and directed the horse in the direction of Babrichi.

Darina sobbed all the way. The pain and the grief she felt from losing her only daughter were just too hard to bear. Knowing that she could never see her child again would make life unbearable and impossible. She had lost the will to live. She wanted to die. Her husband was going to pay them back for what they had done to their daughter without any fear or hesitation. He would fight to the end. She knew her husband. He was a man of his word.

CHAPTER 5

The noises coming from outside woke up the starosta. In a moment he could hear gunshots and the sound of broken windows glass. He crawled out of his bed, afraid for his life. He could almost feel the cold hands of death squeezing him by his throat, drawing close to him. Lying under the bed he could hear someone's voice: "You are rotten bastard! It's your fault that I lost my little angel—my pride, my joy, my only child. She is gone, and you will pay for this, you rotter!"

The starosta knew now who it was and why he was there. He was hoping the SS would hear the shots and come to his rescue. He was shaking vigorously, fearing for his safety. He knew what Pavlo was capable of.

Pavlo galloped the horse from one side of the house to the other. He continued to fire shots until all the windows were smashed and there was not an unbroken window left. "I will make you live in fear for the rest of your days!" he shouted. "You hear me!" His deep voice sent echoes through the starosta's home. It made the starosta's skin crawl and sent shivers down his spine. He was dreading the next moment. "Rotten traitor!" Pavlo shouted.

For the moment it was quiet again. The starosta was too scared to even make a slightest movement. After some time, however, he climbed out from underneath his bed, leaving his wife still cowering there. Making very small movements, he crawled to get his gun. He carefully picked up the gun and started to move very cautiously. He did not want to draw much attention to himself; he was too scared that

Pavlo might see him. He slowly crept to the window. As he tried to look out, a sudden shot hit him in the hand. His gun fell to the floor. Very slowly he slipped down the wall onto the floor. He was in agony. The pain was severe pain as the blood poured out of the wound. Pressing on the area didn't help much. It was pointless to try to stop the flow. He could hear Pavlo shouting.

"I got you, didn't I, arshole? Next time it will be a lot worse! I will let you live, rotten bastard, in agony. Watch your back. Even God won't help you now!"

Shaking and covered in blood, the starosta was horrified. He heard the sound of galloping, but this time it sounded as if Pavlo was leaving. He was holding onto his hand. The blood was running down his arm making a big puddle on the floor. "Bring me some towels!" the starosta shouted to his wife, who was still under the bed. "Quick, woman! Hurry up! I'm dying!" Too concerned for his safety, he shouted in a squeaky voice, "I'm losing too much blood!" He sounded like some kind of creature, too desperate to survive.

The starosta's wife roused herself and tore pieces of fabric. She tried to bandage his hand. She loved her husband no matter what. It was evident she was shaking, but she maintained focus. She was determined to cover the wound and stop the blood from pouring out. She needed to save his life. "I told you not to volunteer to do this job," she remonstrated. "Look at our children!" She was crying hard now. Their three children were terrified. They had come out of their rooms and were sitting tightly together in a corner, their eyes full of fear.

"Shut up!" the starosta said. The blood was still gushing out from his wound.

"What do you know?" She spoke through her teeth. She was very tempted to leave him, but her decency made her get on with her duty and save his life.

"Just fucking get on with it!" he shouted in a squeamish tone of voice. "I'm losing too much blood!" He was repeating the same thing over and over again. He was so afraid. Suddenly he lost consciousness and began to snore and then went quiet. His wife slapped him in an effort to get him to come out of his swoon, but nothing would help. Not even the work of her small hands could help him. He remained unconscious.

CHAPTER 6

Pavlo was galloping to Babrichi. He had done what he had done, and it was not the end. Soon he was inside Maxim's house. Two women were sitting at the table. "Where is Maxim?" he asked.

"He's out," Yarina replied.

"I can't stay," said Pavlo. "The SS will probably look for me after tonight. I shot the starosta. The bullet hit his hand. I wanted to kill the bastard for what he did." He turned to Darina and said, very protectively, "Stay here with Yarina. They will take care of you. Stay strong. I must leave you now. Tell Maxim I will be in the forest near the lake. Tell him to whistle few times, and I will come out."

Darina interrupted Pavlo. "Is this it? Oh, no, no …" She was shaking her head. "You are just going to walk out of that door thinking that I am going to let you do this alone? Oh no, you are not leaving without me. We are leaving together." Darina was a very stubborn and serious woman. She would not leave her husband. In this risky situation, she needed to be by his side supporting him and not leaving him under any circumstances.

"No, stay here, I said." He was firm.

"No way. I am going with you. She also was firm, short, direct, and to the point. "I'm leaving with you whether we live or we die." Her loyalty was not in question. Her husband was everything to her. There was not a sound of hesitation in her voice. Everyone knew she had no second thoughts.

Yarina wasn't going to stop her. She knew she would do exactly the same thing in a similar situation. She would not let Maxim leave without her. "Wait," she said. "I'm going to wrap some food. You are going to need it, my dear. It won't take me too long." She hurried into the kitchen. Then she quickly came out with some goodies wrapped up in a hand-crafted towel.

"Bless you, Yarina," said Darina, hugging her friend with much appreciation. "Thank you for all your help."

"Don't forget to tell Maxim where we are going to stay," Pavlo reminded Yarina as they were leaving the house.

"I'll tell Maxim. Go in peace, my dears." She lifted her right hand crossed herself, giving them her blessing with the holy gesture. "May God be with you now and always." They walked out into the night.

Moments later, Yarina was sitting at the table thinking about Lina as she mended some clothes and waited for Maxim to return. It was very dark outside. Ivan was already in his bed. Deep in her thoughts, she was thinking about the only daughter she had. The tragedy of the war had made Yarina only one of many mothers whose children had been taken. Somehow they would have to get through it. No one knew how everything would turn out, but right now they were ready for anything.

Suddenly the door opened, and Maxim entered the house. "You're here," she said, getting up to help him. He appeared to be in pain. "Pavlo and Darina just left," she said as she helped him take his coat off. "They are staying in the forest, hiding from the SS. They can't stay in the village after today. Pavlo was shooting. He blew all the windows in the starosta's house. No doubt the starosta is going to tell the SS. They will probably send guards to watch over him."

"Yes, but they are not going to guard him for the rest of his life, are they?" he wondered

"As soon as the SS don't need him, they will probably top the bastard anyway, if no one else has done it by then."

"Tomorrow morning," he said, "I'll try to pass a food parcel to Lina again."

"Only God knows how I miss our Lina. I just want to see her." Yarina was crying.

"Me too, me too, my dear wife," Maxim replied, stroking his wife gently on the back.

They held each other tightly, not knowing how their child was doing. Maxim kept talking to his wife holding her tighter and tighter, trying to comfort her. He knew he couldn't do anything else.

"We have to keep strong," he said. "This is what we must do." He was looking into her eyes and holding her hands. "Let's just put our heads down, my old woman, and sleep for a few hours. Come, Mum. I will keep you warm." Maxim often called his wife Mum. It was his way of expressing respect to the mother of his children and the wife he loved with all his heart, dearly and sincerely.

CHAPTER 7

Lina had been receiving her mother's food parcels. The large door would open with the most horrible noise that would wake everyone up. The crowd of captives would stand in complete silence, each listening for his or her name to be called. Although the names were read out in alphabetical order, some people still thought it would be worth hoping; perhaps the name they were waiting for had been missed by the reader.

One morning, a vague unease began to spread among the crowd. One of the captives burst into shrill tears; one of the visitors started shouting out the name of someone who had been waiting for ages to receive some food. And what had until then been just a slight forward movement turned into a huge avalanche as the entire crowd of visitors hurled itself at the doorway through which the food was passed. The SS men who had been standing on guard throughout stepped in and formed a human chain across the entrance.

When the names were shouted out, the parcels would be chucked in. Lina was already waiting to catch her parcel. She knew it had been prepared by her mum, and she knew her dad was bringing it. Lina could taste her mother's special cooking. It smelled so good. No one else could cook like her. When the food was brought in, she shared with Katya and the others. Not a crumb was ever left. Everything was eaten. The people inside were starving; not all of them received parcels.

The girls would often sing to keep their spirits up. It had been so long since they had been brought and left there. They didn't know what was going to happen next, and no one would tell them. The damp walls of the barn were covered in mould, and the smells could turn a person's stomach inside out. It was so unbearable, they couldn't wait for the changes to come. They were all desperate for a bit of fresh air after being stuck inside for so long. It was becoming more unbearable to be in the dingy, dark barn with every passing day. The young people dreamed about the daylight, sky, trees, and the greenery of the countryside. They were tired of seeing grey and mouldy walls day after day after day as they were locked away from freedom. It was all because of the war, and it seemed as if everyone was caught up in it. Deep in her thoughts, Lina turned round and looked at her cousin. Katya was sitting deep in her thoughts too, looking very innocent.

"Remember how we used to play in here?" Lina asked Katya. "We'd run on top of the piles of grain and slide to the bottom. Then we would quickly sweep up before your mother came to check on us. Remember those days?" she said with a sad smile.

They both smiled at the thought of the happiest moments in their lives. Now this place looked nothing like it had looked then. Lina could still remember the fresh smell of the grain and how wonderful it was to play here with her cousin. She had been just a happy child growing up in a good community amongst friendly villagers.

When Lina was sad, her cousin would often hug her making her feel better. No one knew that the war would turn the world into a place where freedom and enjoyment would be robbed from the deserving lives of youth. The world outside was turned completely upside down. They had only each other to hold on to. The memories of their happy childhood and lots of fun helped to take their minds off the torments of war.

The doors suddenly opened up, and another group of people were brought in. Lina recognized her Aunt Dasha.

"Katya, look! It's Dasha!" Lina exclaimed.

Lina and Katya got up and quickly rushed towards their aunt. Dasha noticed the girls and was trying to squeeze between the people to get to them. "Oh, girls, you don't' know how glad I am that we are here together!" she exclaimed. Her eyes were red and puffy from crying. "I knew I was going to see you in here. Your dad was round our house," she said to Lina. "He told us about you. Here is a note from him. Here, in my pocket. Just one second!" She took the note from her jacket and gave it to Lina.

In the note, Lina's dad said how much they missed and loved her. He encouraged her to be strong no matter what. He said that the house was not the same now that she was not in it. Ivan kept asking them when Lina was going to be back. Her dad promised to bring her food daily. He reminded Lina about several things: "Remember, my daughter, don't try to run away or steal! They could shoot you for it. The mission of the SS is to supply young boys and girls to work for them. If you do what they tell you, everything will be fine, and you will be able to return home. Pray, pray and pray, my child! We love you with all our hearts. May God guard you in all you do!"

Lina was very sad after she finished reading the letter. She felt the same pain she had felt when she had been dragged from her house. Gripped from within with emotion, she sobbed. She could not stop the pain as the tears rolled down her face. She wanted to be with her family, not where she was right now. It was so difficult to come to terms with where life was taking her. Who could she blame for this? Why had the war started anyway? She felt only anger and hate. She hated Hitler.

"Lina, come on, darling," said Aunt Dasha. "We are all feeling the same! Let the pain out, and you will be okay again."

In few moments, Lina was calm, and only sudden little sobs overcame her. She felt horribly empty on the inside but glad that her cousin and aunt were here with her.

"You know, girl, I just want to tell you how lucky we are just to be alive," said Aunt Dasha. "Since you've been gone, a lot of things have happened in the village." Dasha began to tell the girls that a lot of youngsters had been shot because they were trying to escape. Some had managed to hide in the forest, and some of those had been found and shot on the spot. "But not all were found!" she said it in a victorious tone. "You know Nina? Pavlo and Darina's daughter? The SS shot her. They didn't even hand the body over to her parents. And why? Because she was ill!"

"Yes," said Lina. "The SS came inside and got her. We were thinking she would be saved."

"But the ugly pigs murdered her and burnt her body!" said Dasha.

"Oh, dear Lord." Lina was speechless and shocked. "How could they have killed the sick girl? She was so helpless! They could have handed her over to her parents."

"Stop it!" Dasha said firmly. "It would have happened sooner or later. They don't need the sick. They need the strong and healthy."

"We must stop blaming ourselves," said Katya. "There was nothing you could do to save Nina. She was sick. You saw how she was coughing—and the rash that appeared all over her skin."

"No one would have thought that the girl would be killed," Dasha said sharply. The girls' aunt was a bit older than Lina and had a strict character. Somehow she was always right.

"May she rest in peace," Lina said in a whisper.

"I know how you feel, but don't blame yourself for what has happened to Nina," said Dasha. "No one is safe here. The war has caused huge elimination of humanity. The war is the main reason for everything bad we are going through right now. Look at us —all poor souls sitting in dingy places, scared of the SS beasts outside. They control us, and we are too scared to stand up to them. We are waiting to be transported, but no one knows where we are going, and we won't know until we get there. Let's just wait and see," Dasha concluded in firm tone.

"Anyway," she continued, "Here is some news that might make you feel better. Pavlo, Nina's dad, went to the starosta's house and really did some damage. He shot the starosta in the hand, but the man is still alive, the cold hearted bastard!" Dasha exclaimed. "He just came out of a coma after losing much blood. His days are numbered. The fewer people who like him, the better it will be for us.

"Don't speak too loudly," said Katya. "There are some traitors here too. Keep your voices down."

"So where is Pavlo now?" Lina whispered to her aunt.

"Hiding somewhere in the forest. Darina with him. She didn't want to stay with your parents, Lina. She's one of those women who stand up for their men no matter what. Some people in the village say that he's joined the partisans. I'm sure he wanted to kill the starosta, but only wounded him instead. The starosta should stay away. If he returns to the village, the people will simply tear him apart. If they don't, the SS will kill him once they're done with him and his 'good work'. The bastard is too scared to go out of the house now. He knows so many are after his butt. Some of the SS guard him because his life is in danger, but the SS need him only for the time being. Soon this will be his lot," She demonstrated by running her finger across her throat. "Let's stop talking about the dead," she finished.

"It's just that he makes me feel so angry." Lina put her little hands into fists. "I want to see him suffering for a long time after what he did to me."

"I just feel for the parents loosing poor Nina, the only child they had," said Katya.

"Their daughter was the best thing that happened to them," said Dasha. "They could not have children for a long time. They were trying to have children and were blessed with only one. They are a nice couple, and how heartbroken they are now."

The girls were quiet for a moment, and the silence was a sign of their grieving for the girl—a minute of silence for one they knew had been spitefully killed. They were praying for her parents, who had experienced shockingly irreplaceable loss.

"Dasha, go get your stuff and move here where we are," said Lina. By the way, this is Kiril! she said, as the boy approached them.

"Hi. Nice to meet you," Kiril said with a friendly smile.

"Hi, Kiril. Nice to meet you too," Dasha replied.

"Kiril is an orphan," Lina explained. "He's with us. We feel that we need to look after him.

"You are in good hands!" Dasha said with a cheerful tone in her voice.

"I'm just one lucky guy to be around such beautiful girls!" Kiril replied with a nice smile.

Dasha didn't know how to ask, but with growing curiosity she finally took the chance: "If you don't mind me asking, what happened to your parents, Kiril?" Dasha was the kind of person who asked everyone questions and wanted to know everything inside and out.

"My parents died when I was just five years old. They died from typhus. My grandparents were bringing me up until I lost them too. Since then I lived and worked on the farm and earned my own bread. The SS men took me straight from the farm in the middle of my work when I was spreading fertilizer all over the field. I smelt like a skunk, but even that did not stop them from taking me! The SS came with the starosta. They asked my name and ordered me to go on the back of the truck. That is how I ended up here."

Lina was looking at him and feeling sorry for the way life was treating him. He looked so innocent and so undernourished. He was just skin and bones. Lina had been sharing her food with him. Somehow, Kiril reminded Lina of her little brother, and she wanted to look after him. The boy was severely undernourished.

"Okay, guys, let me show you what I have got here!" Dasha unwrapped the paper. She took out few slices of home-baked bread, a few eggs, and a jar of milk. "Take it. Take it!" Dasha offered with sincere generosity putting it all in front of them.

Lina loved her aunt for her amazing generosity as well as her amazingly brave character. Her aunt was only three years older than she. Occasionally she would be bossy, and Lina would get extremely upset with her. But they would soon make up, and everything would become normal again. As they sat together eating quietly, they appreciated how lucky they were to have the food.

It was very quiet. Occasionally they heard sounds of people crying, people talking loudly, and very rarely, people laughing or singing.

The youngsters had been unwashed for weeks. The smell of the place was unbearable. There was no way they could go out of the barn to the toilet, so everyone decided to use one place and cover it with raked earth. The smell was just sickening, but they had no choice but to suffer these horrendous conditions as they hoped for positive changes.

And so, one more day amongst so many others came to an end in the barn. Who knew how many more there were to come?

CHAPTER 8

At night, Lina had lots of disturbing dreams. She dreamt about her brother. She saw his face and his eyes full of tears. He was calling for her. She wanted to go to him, but she couldn't because someone was holding her and wouldn't let her go. It was a man wearing a grey-green SS uniform with a rifle slung over his back. He looked dirty from the hours spent on the road. He was bold, and his horrendous face was covered with the scars. She looked down and noticed that her body was wrapped up by a very strong wire. When she looked up to see where her brother was, the image of him has disappeared. She turned around and saw him standing. Lina would wake up with cold drops of sweat all over her forehead. She had these nightmares frequently since she'd been taking from home. Lina's heart ached, and deep sadness inside gave her shivers. She turned around and looked at Katya, who was still asleep curled up like a kitten on her blanket. Lina could not see Dasha anywhere, so she stood up to look around. She couldn't help but notice that something was happening in the barn. Hundreds of worried gazes were looking in the same direction. She tried to see what was happening. She started to walk in the direction where everyone was looking.

Lina noticed a group of people in the corner of the barn, so she hurried to see exactly what it was all about. As she came closer, she was able to see that some of the lads were digging the ground to create a tunnel so that they could escape.

That was it! She couldn't keep quiet. "Have you lost your minds?" she shouted. "What are you doing? Are you trying to get us all killed?"

She pulled one lad to the side. She was strong for a young woman. No one would mess about with her. "Listen, why don't you and your friends just stop? Stop it boys!" she said. "The SS will shoot you on the spot. You know this, don't you?" she said it raising her eyebrows.

Lina began to cry now, the tears streaming down her face. She was concerned not just for her own safety but for the safety of all the people trapped in the barn.

Finally, others started to come forward telling the boys to stop what they were doing. The lads came to their senses when they realized that, by escaping, they would be putting everyone's life in danger, and that was not the way to be. Panic and not knowing what was going to happen had made them do outrageous things before they had thought about the consequences.

Putting her hand on Lina's shoulder, Dasha reassured that everything would be fine. They both returned to their area, relieved that the boys had come to their senses.

Katya was still sleeping with a sweet smile not knowing what just happened. Lina and Dasha could not believe how deeply she was sleeping. Suddenly, the thunder roared, and the faces of the terrified girls were lit by lightning flashes. One could tell the rattle of rifle fire from the noise from the heavens, which made Katya jump in her sleep. It rained all night. Not many could sleep. They huddled together, trying to keep warm.

Chilled to the bone, hungry, and tired, Lina sat between her cousin and aunt, trying to keep warm. They hadn't eaten since the previous day. Everyone was wondering what was going to happen to them. They had passed another day; they didn't know what the next day could bring.

PART 3

DEPORTATION

CHAPTER 9

A loud siren disturbed the silence of the night and announced the morning hour. The people in the barn could hear the noises of the squeaky railway wagon wheels. The old train had stopped nearby. They heard the footsteps of the SS men heading towards the barn. What happened next filled everyone with dread.

In one moment, the doors opened wide. The SS men ordered everyone to get up, take their belongings, join the queue, and come outside in single file. The soldiers shouted orders, marshalled everyone into lines, and prodded everyone forward with their gun muzzles. Some of the captives were crying in fear. Lina was nearly in tears herself.

"Schnell! Schnell!" (Fast! Fast!) Two SS men, one on each side, were pushing people one by one. The both had disgusted looks on their faces, as if they were herding cattle. "Schmutzigen Schweine!" (Dirty pigs!) They were laughing in satanic tones as they twitched their faces and spit, revealing their filthy yellow teeth. They didn't care about anyone. Their prime concern was to transport people and exploit them, ripping families apart, causing heartache, pain, and human tragedy without a second thought. They were ill minded, having been corrupted by the Hitler's beliefs.

Some of the captives fell in the crowd, not so sure whether to go left or right. People got confused about the rows, each person desperately seeking to survive. No one knew who would stay alive. Those who wanted to pee did so while walking along, the warm liquid soaking into trousers and rapidly cooling. Those who wanted to poo … didn't want to poo. Some people began to push so hard they caused others to trip. They landed with loud thuds, scratching their skin on the uneven ground that was covered with pebbles and debris. This repetitive rhythm was punctuated now and then by the crack of a rifle as another heavy bullet, fired at close range, slammed into a fragile body. No bullet ever missed. No target ever survived. They all had to keep their heads down so they would not become the next target.

Taking advantage of the sudden chaos initiated by the SS shooting, Lina unfastened the golden locket her parents had given her. She knew the SS intentions, and she would rather hide it away than give it to them. Being right at the back of the queue made it very convenient. Something she couldn't even understand prompted her to do this quickly. She was glad to have time to do it. To her surprise, there was already a small crack in the corner of one of the supporting beams in the barn. She quickly put her locket inside and covered it with the earth. It was now in a place that no one knew about. It was her secret.

She joined the rest of the people as they moved to the main doors of the barn. First went the boys and then the girls. One by one they walked in fear-driven silence. The SS men herded them all together, hundreds of them. Black-coated SS men stepped forward and started sorting people. Young, healthy youngsters were put into one group, designated A. Others were allocated to group B. Group A was ordered to set off to the station at a run. Before they got there, they could hear the SS men shooting everyone they had left behind. Those who had survived were shoved violently onto the waiting train, into train wagons that were generally used for transferring cattle. Some of them were beaten in the process. The floors of the wagons were covered with prickly straw and dry weeds. The small uneven gaps between the planks of the walls let a tiny bit of daylight in. The wagons were dark, rough, and smelled of old wood infected by woodworm, rain, and dampness. The captives who were already on the train showed signs of desperation. They were terrified and suffered from dehydration and hunger. God alone knew how long they'd been on the rails. The last thing the SS wanted was a running fight on their hands.

When the train started to move from their village and a whistle announced its departure, the screams and the cries of people made the situation unbearable. Hundreds were shouting, screaming, crying, cursing, and praying. It was extremely touching, this moment of terror. The young lives of the captives were under scrutiny. Parents didn't know where their children were being taken and how they were going to get through this horrific ordeal. Each boy and girl on that train was a captive of the cold-blooded beasts, sadists, crooked invaders whose motive was to destroy the Soviet nation, wipe the nations out, and rob their countries in the most distractive and eliminative war. This was a moment that would leave footsteps in history for many generations to come.

To the astonishment of Lina, Katya, and Dasha, the frail boy Kiril was on the train with them He was so helpless, so undernourished, thin and fragile, but he had managed to survive the "quick run" game played by the SS men. He looked completely confused and didn't know what was going on or where he was. The people were afraid that the SS men would shoot him as soon as they saw him because of his very thin and helpless-looking body. They decided to hide him from the SS. There was something about him. His innocence exposed his vulnerabilities. Every time the wagon stopped the people made him hide in the corner of the wagon under a heap of straw.

The train passed many stations, and the wagons stopped from time to time. There would be a weird silence punctuated now and again by the snort of the locomotive or the bark of a dog. At some stations people ran towards the train, begging for some food in exchange for jewellery such as broaches, necklaces, and fancy combs. They were crying out for help, hungry, exhausted, and severely undernourished. "Here, here! Take the jewellery!" It was no longer a treasure in the time of war.

Quickly Lina broke the bread into small pieces so that she could pass it to as many outstretched hands as she could. Lina's hands were trembling as she was pushed them through the gaps in the boards. The rough wood scratched her arms and hands, giving her splinters that rubbed and dug into her skin, but that didn't stop her from giving. Her hands bled, but she was ready to give as much as she could. She had to make herself stop and not give away all their food. They must have some left for themselves. "Some more bread. Take it. Take it," she said in a commanding tone of voice. "Look at these poor people. Skin and bones!" she cried out.

"We don't have much left," one of the girls shouted out through the boards. They wanted to help with the food they had, but they knew it wouldn't be possible to feed them all. They didn't have much themselves, and soon they too would be starving. But they gave, knowing that, for many, the small amount of food meant another day of survival, another day of getting through, another life prolonged.

In Lina's eyes, she was choosing life over death, and even a bit of food meant life prolonged for one more day.

Suddenly, Lina saw a woman who was trying to put her hand through the boards of the train. Her hand was covered in blood from pushing it so hard through the rough boards. She was asking for food. With her other hand she was holding a baby wrapped up in the blanket. "Please, can you give me something to eat, dear!" the woman cried. "I need to feed my baby. I need to feed my baby," she repeated this over and over again. "The milk from my breasts has vanished. Nothing left. Not a drop. I haven't eaten for days now."

Lina picked up a piece of bread and a cube of sugar, but then she saw the woman's baby. The baby wasn't breathing. The baby had died, and the woman was still holding onto it as if it was alive. The train started to move.

Disturbed by the scene, Lina shouted, "Pray! Keep on praying!" The woman started to run, still holding onto the boards of the wagon. It made Lina's heart ache to see the horrors of the harsh realities.

"Bless you! Bless you!" the woman was saying as she still tried to run after the train. Lina wanted to shout out to tell her that her baby wasn't breathing, that she should check the baby. But the words got stuck in her mouth. She couldn't say a word; the words wouldn't come out. "Bless you, bless you!" The echo of the woman's voice was still in Lina's ears even though the train had already left the station a while ago.

CHAPTER 10

Lina was sitting on the straw quietly thinking of her home, her parents, her brother, all of whom she loved with all her heart.

"Are you all right, Lina?" Katya asked

"No, not really."

"Did you see all these people?"

"They were so fragile, thin, and lifeless. Starving people. This war—families are torn apart. Why?" Lina asked. She was sobbing hard. The tears were rolling down her face. Her chest shook with each sob.

Katya couldn't watch her cousin being so upset. She sat beside Lina on the floor and hugged her very tight. She wanted her cousin not to hurt. "I will never leave you, I promise."

"We must be strong for one another," said Lina.

"One day," said Katya, "everything is going to change. We must live and hope!" The cousins looked at each other, still smiling through the painful moment of their war-torn lives.

"Millions joined the army!" said Lina, reassuring her cousin. "They will fight for us, and one day we will be freed. We cannot be in captivity like this forever, you know." The girls kept comforting each other, and somehow it made them feel at ease. "We must believe in better times," said Lina. "Somewhere, somehow, someone is thinking about us. Otherwise we shall all fade away and die ..."

"The men—your dad, my dad—they will fight for us till the end," said Katya.

"Our leader Stalin will never let his people down," assured Lina.

"You know this, Lina," Katya said with so much enthusiasm. Her face expressed so much determination and positivity that Lina felt reassured and stopped crying. "Stalin is a powerful, ruthless leader," said Katya. "He is capable of fighting and freeing our country from the invaders." Katya was trying to give her cousin the mental strength she needed to believe that Stalin could help them. "One day it will happen, but we don't know when." She raised her eyebrows, not having a clue when the day would come, or if would come at all. "You're going to see the day when we will be free," she said, not wanting her cousin to think otherwise.

CHAPTER 11

Lina noticed that one of the girls they'd become friendly with, Marisya, had been quiet for days. They could tell she was thinking about something. She didn't talk much. She was in her little world, in deep thought. But then, to everyone's astonishment, she broke out with surprising news. She talked to everyone about escaping, and she urged other girls in the wagon to follow her lead. The best way to escape, she explained, would be to jump down from the wagon when the train slowed down. It must happen when the train turned a corner. That was the best option for jumping off. "We should do it now, girls," she urged, "whilst we are still in the Ukraine." She kept on and on about it. "Then the best plan for us is to join the partisans. They need more people. My dad mentioned this to me. Who is up for it?"

When no one seemed enthusiastic, she continued, "Come on!" She urged them with so much desire and determination. Her facial expressions indicated how determined she was to undertake such a risky operation no matter the costs or the consequences. By looking at her, anyone could tell she had made her mind up. The only thought occupying her mind was to escape. "I know the SS will eventually kill us all," she proclaimed. "The wild animals' orders came with sticks and rifles. Didn't you notice when a few went down in the crowds? I didn't see them getting up again," she urged. "Now is the best time to escape, girls. It's not going to get easier. The wagons are old, and it's not hard to remove few boards. We can handle it. It's now or never! What do you say? Lina? Katya? Come one, let's do it!" The sound of desperation in her high-pitched voice conveyed her nervousness. Somehow it sent everybody into a state of panic.

Lina was sitting on a pile of straw listening to Marisya. She could really see her friend's determination. Marisya's attitude and very strong ego surprised the girls, and at the same time they couldn't believe how risky her idea was. They knew that nothing would stop her from doing what she had planned in her mind. She wanted to escape even though it was risky. Lina was angry with her because she kept on and on about this foolish escape. It was like listening to a broken record. "Look, Marisya, you want freedom? You are not going to get freedom in this war, whatever you do, even if you escape now! Don't you realize that our lives are never going to be the same because of the stupid war!" Lina was shouting now. "If you escape now from this train you are risking your life, darling!" She was speaking quietly now, trying to reassure her rather than shout at her.

"Stay with us, Marisya," another girl said. And others urged her—almost begged her—to stay.

"Don't do it, Marisya," Dasha insisted with tears in her eyes, her stern look conveying the seriousness of her warning to the girl. "You might escape one captivity and get yourself into another. The SS are everywhere. Nazis spread like a bad disease. Didn't you observe what they did to our village from the moment of their arrival? They are brainwashed by the Hitler's sick-minded mentality. Their goal is to destroy humanity, rip families apart, and cause the kind of devastation that sending chills down people's spines. They have the authority to commit mass murders without being held responsible. Everybody witnessed their announcement of the fact."

"Stay here, Marisya," pleaded Lina. "It's the only way we can survive, and escaping would give them the excuse to shoot you on the spot. My dad warned me, and I am telling you now that we must stick together!" She conveyed her dad's wisdom. "It's either all of us jump out or all of us stay here, Marisya! One of us cannot go alone," Lina emphasized her point letting her know she was deadly serious about what she had to say. "Besides, the SS men watch all our movements," she said, stating the obvious. It couldn't be clearer than that. "We cannot risk broken bones that would threaten our survival. Also, we all know Kiril is too weak and could not handle an escape. He is just not fit to be part of the operation you are suggesting. It would finish him, and leaving him is out of the question. This would kill the poor bugger. Don't you get it?" She firmly stated her concerns, hoping that Marisya would come to her senses and change her mind. "I feel obliged with a duty to keep him protected from the enemy's cold-blooded devilish attacks. If we left him, we would be no better than the SS men with their sick mentality. Nazis think they can rule, but they won't be in power for long though. Sooner or later, the world's armies will get hold of them and blow them away."

No one else was speaking, but they all nodded; they all agreed with what Lina was saying. She had spoken clearly without a bit of hesitation, but with pure determination as she tried to talk sense to Marisya.

Still, the strong opinion of her companions did not convince Marisya. She was so stubborn, like a donkey. She was going to do what she had described. Because of her stubborn character, it was very difficult to make her change her mind. The girls knew her too well. Once she made a decision, there was nothing that could stop her. The determined Marisya sat next to the sliding doors of the wagon. Every time Lina woke up from her sleep, she saw the girl sitting there. Her silhouette rocked from side to side with the rhythmical motion of the moving train. In the morning, as the train slowed down, Lina woke up at the loud sound of shooting. Marisya was no longer in the wagon. She had done what she had promised to do.

Dasha shook her head in disapproval. "How could she do this, silly girl!" She still couldn't believe the actions the girl had taken even after they all had tried to dissuade her, tried to talk sense into her. "She could be lying in a ditch with broken bones for all we know." The others were numb and speechless with nothing to say. "She's just one of those people who can't be stopped once they make up their minds. Not even close friends could stop her. It's as if she was blind and put herself into a deadly position—something she couldn't even get out of. Her mind was her prison. She thought, she did."

"Yeh, you are right," Lina said, nodding.

"When you think about it," said Katya, "she could have stayed here, but then everything happens for a reason."

"We tried to tell her," said Lina, "but there was nothing we could do to convince her to change her mind. It was too late and pointless. It was pointless to try to reassure someone who had made up her mind." They had tried to talk her out of it, but it hadn't worked. All they could hope was that she had been able to stay alive.

The SS men watched the girls' every move from then on. The girls sat in the wagon like mice, so quiet one could hear the fly buzzing in the air. The slightest attempt to escape could determine their fate there and then. They didn't make any attempts; it would have been pointless. Lina knew her father had been right when he'd said, "Don't run, because they can kill you." The Nazis used various tactics, and intimidation was the name of the game. The captives who stayed in the wagons were prepared for anything. They didn't have the physical or emotional strength to fight or to take on an army of SS men with dogs, guns, and severe torture techniques. In fact, these combined techniques were part of the most sophisticated tortures dreamed up by the SS. This was actually brilliant psychology. The SS were obsessed with things like regimentation, instant blind obedience, and mindless subservience. The heel clicking militarism echoed a sinister, appalling purpose, a savage adherence to the sick mentality of the Third Reich.

This made captives too afraid to look up, to make slightest attempt to express concerns, or to show any signs of anything out of ordinary, for any movement could draw the attention of the SS and make things worse.

CHAPTER 12

The train picked up the speed again. Only the noise of the moving wagons disturbed the silent night outside. Suddenly a loud clap thunder was followed by lightening that lit up the sky, making silhouettes of trees visible through the gaps in the sides of the wagon. The trees looked like lonely soldiers standing at their post too scared to move. There was another rumble of thunder directly overhead, and the sky once more crackled with lightning. In minutes, the sky turned dark blue, almost black. The raindrops hit the roof of the wagon, seeped through the gaps, fell onto Lina's face, and ran down onto her clothes. The thunder rolled, and the entire interior of the wagon was lit by the lightning flashes. Then the rain poured down from above sending cold chills through the captives. Lina tasted the rainwater on her lips. The rattle of thunder reminded them of the roar of rifle fire and made them all jump every time they heard it. Burying herself underneath the straw, Lina tried to shelter herself from the rain that was seeping through to her skin. She was trying to think of all the good things—her home and the dear people in her life. She thought about her dad, who was so caring. As long as she could remember, he had been there for her—he who would forgive her anything. Her mum was her strength. She had taught her so many things. Lina's memory drew her back to the times when her twin sisters were alive and how all of them together would chase Ivan near the river embankment, playing and laughing and threatening to throw him into the river. He was always so fast on his feet. No one could run faster than he could. No one could ever catch him. He was one of the fastest boys in the village.

Lina thought about her mum. All the women in the village used to get together on a Sunday. In fine weather, they would sit near the river embankment and talk about all sorts of things—recipes, cooking, and children. The little ones would swim in the river, giggling, throwing each other, jumping from the little wooden bridge, diving into the clear water dappled with the reflections of the sun. From time to time, Lina envisioned her mum's smiling face. In her mind, that was a world beyond what it was now where everything was perfect and without distortion or imperfection, a world where the children formed character with complete truthfulness and rattled off their verses without stumbling or hesitating in the slightest. They had experienced nature's greatest gifts. Everyone had observed the pleasant surroundings, even noticing the birds spying for fish so they could get their catch of the day. The children's loud laughter sent echoing sounds as they ploughed their way through the water.

These moments were stuck in her mind forever. That had been a time of peace, joy, and childhood freedom and innocence. Every time she thought about those times, those happy times, she felt at ease and was able to forget for a brief time about the war. She hated what she could see now. The triumphant times had been short lived, and she drifted off to sleep again. She dreamt about her home, the freshness of the morning meadows, the sound of singing birds, and the tranquillity of her entire motherland.

Suddenly her dream changed. She saw her parents' home. Ivan was sitting on the doorstep. His face was dirty from the coal. His clothes were covered in dust. The palms of his hands were firmly pressed to his jaws. The front door opened. She walked into their cottage, but there was no one in the room. She searched for her parents, calling their names, but she couldn't find them. She ran outside, and Ivan was not there anymore. She wanted to call for them but she couldn't. She was struggling to find them. She couldn't find them anywhere.

CHAPTER 13

Lina woke up, drops of cold sweat covering her face. She wiped her cheeks with palms of her hands. Why did she have that nightmare—that same dream—over and over again? It was a very unpleasant dream. She got up and tried to look through the gaps in the side of the wagon to see where they were. She did not want to think about the horrible nightmare. It had rained all night, and she felt chilled to the bone. She realized that the train was going through a station in Poland. She knew this by the writing on the worn-out wooden signs.

The train did not stop for days. The girls were cold, hungry, and thirsty. They felt weaker and weaker with each passing day. From time to time, they would be given some water and dried bread. The doors would fly open when the train stopped at a station, and the food would be thrown in as if to some animals.

Dasha grabbed onto the piece of dried bread and held it with such greed. Her hands were trembling, but she found it hard to put it directly to her mouth. The inside of her cheek was bleeding from ulcers, and the dry bread made the little wounds inside her mouth bleed. It was so painful, she grabbed onto her cheek. Agony showed on her face. She started to cry. It was so painful she couldn't take it anymore.

"Here, have a sip of water," said Lina. Dasha washed her mouth, spitting the water out. It had soothed and relieved her symptoms. The hunger was too much to bear. Despite the bleeding and the pain, she still tried to eat the dry bread by sucking it. It was the only option they had.

The train finally arrived at their destination. They were in Germany now. The train had been moving slowly now, stopping only occasionally.

Another stop, and this time everyone was ordered to leave the wagons and come outside. Everyone huddled together shocked, dazed, and unable to walk quickly. If someone wasn't moving fast enough the SS dogs were commanded to bite. Many people were knocked off their feet, knocked to the ground.

Everyone was taken inside an old-looking building where they were ordered to strip out of their clothes. The SS doctor examined the young men and women. They were ordered to line up, and then the line was divided further to the left and to the right. Not one knew what was going to happen to them. In dread, they shook from the cold air inside the building and from the stern, angry looks from the SS men. Then the captives were ordered to get dressed and return to their wagons. The examination had been completed. They were in despair that was beyond any description, but they somehow maintained a notion of survival, the will to go on, day by day, coping with whatever the Nazi bastards threw at them.

"Schnell! Schnell!" The SS men were shouting, the foam and the spit coming out from their mouths as they hurried the captives along. They always treated the youngsters with shouting and hitting.

The cries in the crowd made the enemy more aggravated, and the shooting scared everyone even more so. A few in the crowd were shot. There was nothing the captives could do to change anything. They had been taken. The war was taking them in an unknown direction. The greedy SS had taken all their dreams and youth. Deprived of the most precious of gifts, no one knew who would survive, who would return, and where all of them would end up. Some of them would die from malnutrition, exhaustion, and disease. Others would go down under the boots and gadgets of the Nazis. The rest would be escorted to places to carry out further work orders. The Nazis were taking the youngsters to the gates of death. All would suffer. Some would starve to death. All would be abused. Many would be thrown into a *bratskaya mogila*—a common grave—with no mention of their names. They would have paid the price for crimes they never committed.

They were being ordered to get inside the train once again, a train that would take further and further away from their homeland and into a foreign land.

Another stop. The train pulled into the station with a screech of brakes. The wagon doors were thrown open. "Raus! Raus!" (Out! Out!) They heard the shouting.

They didn't appreciate how awful and heartless the SS language sounded. There was a wall of noise as a voice snarled orders over a loudspeaker system. The SS men were working now, carrying out their duties with the efficiency for which Nazis were famous.

The girls quickly hid Kiril in the straw again. He was the most vulnerable. He couldn't even stand up properly without someone holding him. The SS would take him away, and the girls knew what that meant. They were hoping to save him. They followed the commands, jumping from the wagon walking over the rails to the place where everyone was lining up. Lina's heart was pounding. She could feel it beating in her chest as if it was going to pop out. She didn't know what would happen to her and the rest of the young people. Where they were going? What was going to happen to them?

She watched the Waffen SS men pushing and prodding with their guns. They went into the wagon and dragged fragile Kiril outside. His face with visibly thin, his skin tight over his features—a sharp nose and a thin neck. His narrow lips were moving slowly, exposing ivory-coloured teeth affected by the tooth decay. He was ordered to go to the left with the boys. The girls were ordered to go to the right. The SS men pointed with a stick, gesticulating as they verbalized the orders. They indicated who had to go to the right row and who had to go to the left row. The girls didn't even had the chance to say goodbye to Kiril—there was no time to say anything. Lina could no longer see him in the rows of people. The captives stood in rows behind the SS men who stood guard with their guns and dogs while other trains pulled into position and waited. The SS men went about their business, checking that everything was ready.

Once again the order was to return to the wagon. Kiril did not return to the wagon. The girls were all numb with shock. They sat in the corner of the wagon crying and praying for his safety. He was so thin and helpless that he would not survive the kind of blow that SS men so enjoyed giving out. He was very innocent and looked as if he needed some protection. This was how Lina felt.

Suddenly, as if prompted by some invisible force, Lina began to sing. All the girls started to sing along at the tops of their voices. It was the only way to keep calm. The haunted, sad faces with tear-stained cheeks, had panic written all over them. The girls were trembling, clustered together, but they sang to calm their anxieties. The sound of the train wheels joined the rhythm of the music, creating an everlasting flawless melody that sent echoes into the dark, deep night outside with only occasional disturbance coming from the noisy wagon wheels. The train was taking them further and further away into the unknown.

The train arrived late at night. The SS men didn't turn on any lights. It was not safe, as the planes were dropping bombs on illuminated targets. The SS men and their dogs surrounded the train and shouted, "Raus! Raus! Raus!"

Grey-uniformed SS men were there with their lighting flash collar emblems and the smirking death's heads above the peaks of their caps. They had their own German police, the Gestapo, and the ghetto police ran at their heels with their vicious dogs, trying to look significant. The barking of dogs and the clatter of boots echoed around, ringing in the ears of the captives and rumbling around the hovels they would have to call home at least for a while.

The girls were herded into the hovels as if they were a flock of sheep. They were ordered to strip off their clothes again. Lina was a very shy young woman; she was even embarrassed to take her clothes off in front of her mother, not to mention the strange men. Here she had no choice but to follow the order. She tried to cover her body, but it was impossible. She had only two bare hands to cover it with. All were ordered to go into another room. As they walked along, the SS officers smacked each of them on the bottom, laughing with hysterically perverted rudeness.

It was humiliating, degrading, and inhumane, something that the young women had never had to deal with before. In the next room, she saw how girls lost their long, beautiful hair. It was completely cut off, and then their heads were shaved. Lina knew she was going to lose her dark, chestnut-coloured, long, curly hair too. When it was chopped off, she felt sick. It turned her stomach inside out, making her throw up. The SS men turned around and ordered Lina to clean it up. She bent down, sobbing, and cleaned the floor with handfuls of her own precious hair, which had selfishly been chopped off by the death beasts. The SS men laughed and pointed their fingers at her, spitting, and twitching their faces as if she wasn't a person.

All the girls were given uniforms. Each uniform bore a patch that bore the letters OST, which stood for *Ostarbeiter*, a worker from the east. The night was agonizing for all of them as they sat on their suitcases or lay where they could find the space. They were tense, scared, defenceless, and feeling completely numb. They were the "healthy ones", and they had shivered when the SS man in the grey uniform had taken their numbers and written them down in his ledger, part of the ton of paperwork of which the Third Reich was still, at this time, so proud because it documented the timetable of liquidation.

Awakened by desperately wanting to go to the toilet, Lina tried to find a spot where no one could see her. She decided to move quickly before the SS guards could spot her. Desperately, she emptied her bladder in the corner. One of the SS guards pointed his gun. She lifted her arms and went back without giving him any reason to pull the trigger. She adhered to the bastard's sick mentality so she would not get shot on the spot. They did it without giving any warnings. No one had the physical or emotional strength to stand up to or to fight an army of SS men armed with dogs, guns, and cudgels.

One of the girls came up with an idea. The hovels had two floors, so she suggested that they empty their bowels in the corner of the top floor. The place smelled, and it was revolting. On the second floor, they spotted some soil. They did the best they could to keep clean. This was the only place where they could do it. Lina was in this building for two very long dreadful weeks as they waited for all the others to be brought to the same place. Every day more and more young people were brought in from different parts of the Ukraine. The number of people in the barracks grew each day. They hadn't heard any orders for many days and nights. The air inside was very heavy. It was smelly and made them want to vomit. Everyone was hungry. No food was brought in to them. For them, it was another lockdown. The barracks were padlocked, but they could see what was happening before they were all inside. Some would go to the big post in the middle of the room. They would hold each other hands and pray for survival. Lina would occasionally join in, although she found it difficult to find the strength to pray. She was exhausted like most of the captives inside. The post in the middle was like a symbolic place of worship, prayer, and togetherness.

"Dasha, come with me?" Lina asked.

"Why, dear? I'm finding hard to lift myself from this floor," she said in rather tiring voice.

"I'll help you." Lina took her aunt's hand and helped her up.

"Why?" asked Dasha

"Because, when you pray, you get filled with the spirit, and this is how you can get better." She tried to reassure Dasha.

"Yeh, let's pray together."

Together, along with the others, they whispered the words of prayer, lifting their souls and heads up. It was the holy moment of truth, the truth of devotional prayers to the highest of all. Girls head hands and looked into each other eyes. They were feeling the strength that was lifting them higher, making them stronger and happier. It was a moment of unification and connecting with God. The transgression of the situation didn't halt their faith or crash their spirits. Their physical bodies may have suffered due to severe undernourishment, but they hadn't turned from the truth.

Food was a major problem. It was so vital it occupied their minds, and the only way to stay focused was to pray for a breakthrough. They prayed together for weeks that turned into months, and then one morning their prayers were eventually answered. The doors opened wide, and people came in with an enormous big pot on wheels. It was hot food, the moment they all had been waiting for.

The breakfast was made from potatoes. The mixture was very watery with only few potatoes floating in it. Lina threw up as soon as she noticed a maggot in her bowl. She couldn't eat it. Her stomach rejected it despite her hunger. Only a bit of skin or fat had been thrown into the grey soup to give it any illusion of flavour. Lina's stomach almost stuck to her back.

The next day, in the early hours of the morning, a girl called Tanya came in. She was the one who was serving the food for the day. She warned the girls not to return the food to the kitchen because that could cause them to lose a piece of bread, a little cube of sugar, and tea. Tanya's message travelled fast as girls whispered to each other so that everyone in the barracks knew about it.

The plan worked. The next morning they were given bread, a small cube of cheese, and tea with sugar. Lina tried to eat her bread as slowly as she could. She chewed slowly and indulged in every moment, not leaving even a tiny little crumb behind. Who knew her life could have changed so dramatically that she could be so hungry for every little crumb of bread? She never knew her life could change so much that she could endure such hunger. Her only desire was to eat. Food was the only thing that occupied her mind. Her curvy body had turned to bones. Starvation had made her even hallucinate about food, as if she could smell the aromatic delicious meals her mum used to prepare. All that was in the past. She was so hungry, her stomach would rumble as if some creature was inside of her sucking it all out.

No bread was given the next day. In vain, they looked out the window at the place where the food came from. They received soup that was just boiled water with a few grains floating in it, making it beige in colour. It looked and smelled just like ammonia. It wasn't appetizing—far from it. However, everyone fell on the food, even those who said earlier that they wouldn't touch it! It turned out to be the only meal of the day. The smell of the soup stayed in the walls far into the night, and it was like the ache in their stomachs tightening everything inside. *Will I ever get used to it?* Lina thought to herself. Now she could only dream about the soup she used to eat.

As new people arrived, the youngsters tried to find spaces to sleep. There was so little space, one person said, "Make your bed on the floor. Take the rug out, put it on the floor."

Another said, "Lucky we have the straw here. Otherwise we would be on the cold concrete floor."

Some lay on the floor staring into the empty space. There was nowhere else to look. It was as if they were waiting for something, but no one knew their future. No one knew what turn it could take. No one knew when they were going to be fed again. They were packed like sardines waiting for the selections to start.

In the evening, when no one came into the barracks, the last hope for bread disappeared. The captives had been left to starve again. Undescribed weakness had made it hard not only to walk, but to talk. Thoughts of death slowly started to creep inside their minds. The hunger, the starvation, had made them loose vitality and had drained their intellect and spirit.

"I can't die like this!" Lina spoke her thoughts out loud. "I think we have to put all the food in one pot and share amongst us, even if it's just a bite or two. We must encourage each other. It seems the only way to stay alive."

"That's right—eating often but very little," Dasha agreed.

"We can only have one tiny piece of cheese and a tiny piece of bread," said Lina. "We are running out."

"How horrendous," said Dasha. "I have a migraine. I feel as if a tight band is wrapped around my forehead. The hunger is killing me."

"I'm thinking about nothing but food," Lina confessed. "I think I'm going crazy. Food occupies my mind day and night. It is as if hunger has got its teeth into me, and I am being turned inside out. Every bit of my body is crying out for food; anything will do as long as it can be chewed and swallowed and taken down into my stomach."

"No, it's on my mind too," Dasha replied. "They should bring the food in tomorrow," she said, stretching her hands out in her yearning.

"We have to pray and hope that all of this will soon change," said Lina. "Otherwise, no one is going to make it." She spoke in a quiet voice, too afraid someone might hear her. She was very private and didn't like to say too much, but sometimes she would speak out when she needed someone to hear.

Not everyone was thinking along the same lines. Everyone fought for survival in her own way. Sitting on the floor cross legged, full of fear and suffering with malnutrition, they were waiting for something to happen. They counted days, weeks, and months.

The situation started to change when young men and women were selected for work. The owners of the factories, mills, shops, and farms arrived daily, selecting only the fit and able. Once a person was selected, it was the start of a hard, long, laborious, agonizing journey to the unknown. The war had got them into the situation they were in, and there was nothing they could do to change that. Not for now, anyway. Something had made them want to keep going. In the locked-up barracks, the only progress was by means of selection—somebody else's choice. It was like the worst kind of nightmarish interview because the captives had no idea what the selectors wanted. The captives could only hope for the best.

PART 4

SLAVERY IN GERMANY

CHAPTER 14

It was selection day. All the young women had been called to the place where the selections were made. The room was cold. The concrete walls were grey, and the darkness and the stale smell in the heavy air indicated that the windows were shut. Who knew when someone had opened them last? In the centre of the room was a large table with chairs around it. Nothing else was in the room. The place was empty, cold, and uninviting,

The selected group of young women, one by one, stepped into the room. They were of medium build with relatively good health records. This was enough to be selected for work as labourers. Ten in total were brought into the room. They stood in front of a hard-looking SS officer with ginger hair and moustache and a strong physique. He was stern looking, and Lina's observations sent chills through her body. She didn't like him. There was something very annoying and unpleasant about the man in the uniform. She was not looking forward to the next step because she sensed it was going to be very rough and unpleasant. The anger was in evidence all over his face. He was sweating, and the wide pores and pimples on his skin and the greed in his eyes made him look like a wild animal searching for his prey.

The young women were standing in a row trembling, not knowing what was in his head. He walked down this line, lifting their heads and giving them an unpleasantly stern facial expression. He smelled of sweat. His big nose was covered in pimples and looked like an out-of-shape potato. It was enough to make anyone throw up. When he touched Lina's chin, she prayed for him to leave her alone. The unpleasant smell hit her nostrils, and all she wanted to do was throw up. She was trying to keep it together. Finally, he moved to the next girl. Lina was relieved he had left her alone.

Out of ten, seven were selected and taken to the next room where a doctor examined each girl, checking her hearing, eyesight, and making sure she was fit and well enough to leave the hovels and start working.

"What place in the Ukraine you came from?" an SS soldier asked the related young women, Lina, Katya, and Dasha. All three girls were standing next to each other too scared to be separated.

"Zsitomir." Katya was the bravest of the three.

The door opened and a man in his late fifties walked into the room. "*Dobriy den!*" he said in broken Ukrainian. "My name is Alfred." He smiled at them. "Where are you from in the Ukraine?"

"Zsitomir," Lina replied.

"What are your names?"

Lina, Katya, Dasha gave him their names, one by one.

The man selected Lina and Katya. The girls seemed a bit friendlier than their Aunt Dasha, and he thought it would work out much better for him, knowing his wife's character. Dasha was a very serious young woman, and he was put off because he was thinking about pleasing Anna.

Dasha was selected to work in the village factory.

The girls were taken into village called Kemarchaem. All three were ordered to work for different people. They were sad at being split up, but they did not lose hope that they would see each other again. This somehow gave all of them a sense of courage.

Lina arrived at a very nice big house with a lot of land around it. The property surely required a lot of work. It was evident that the owners took good care of the landscaped area land around the house. She saw neat rows of growing vegetables. She even looked forward to starting work now.

A woman in her early fifties came out of the house to see the new arrival. Lina guessed it was Alfred's wife. It was evident by the way they exchanged their gazes. "Hello." Lina spoke first, trying to make a good impression.

"Hello," replied owner's wife with a surprised look on her face. "My name is Anna, and yours is?" She squinted her eyes as if trying to make Lina hurry with the answer.

"Lina," she answered quickly.

"Welcome to our house, Lina." Anna invited her inside, opening the door and letting her in without any hesitation. She had been expecting the new arrival, and she had prearranged everything and planned well. They had prepared a room for the young lady worker.

First, however, Anna ordered Lina to follow her straight into the bathroom and told her to wash thoroughly, showing her what she meant with the hand gestures. Anna gave Lina clean clothes. "Your old clothes are going to be burnt," she explained, and told Lina about the head lice epidemic.

There was another epidemic that absolutely terrified the SS. It was typhus. Many had lost their lives. The SS had developed tactics to deal with it. The disease was spreading fast. First, they burned all the clothing. They also used a chemical as a disinfect and insecticide, but most often the substance was deadly. It was called Zyklon B.

"You are going to eat with us at the same table," Anna pointed out. "You must learn German." Lina watched the way the woman's thin lips threw the words out—words she couldn't understand. "You must speak only in German to us ... *verstehen*" (understand), Anna insisted. "I don't know Ukrainian, and besides I am not interested in learning your language." Anna seemed a little annoyed as she showed a fake smile and uneven teeth. "Is this clear?" She gave Lina a stern gaze and raised her neatly plucked eyebrows.

"Ya, frau Anna," Lina politely replied, holding her head up.

There was something about Alfred's wife. Lina wanted to like her, but the commands and the way she chose to look—her squinting eyes and some of her facial expressions—left her indecisive in the way she was feeling about the woman. The pitch of her voice was squeaky, and it was off putting. Still it was very early days for forming an opinion about the couple.

After dinner was over, Lina was taken to her room. It was on the second floor in the big house. They walked together up the stairs, and Anna opened the door. It was a small room with an old chest, a small wardrobe, and a small bed. The room had only the necessary items. It was clean and a good place to put one's head down for the night. When the door closed behind her, Lina fell onto her bed. She lay in her room thinking about her home, her parents, and her little brother, wondering how they all were. She loved them with all her heart. Slowly she drifted off to sleep, pulling the quilt to cover her shoulders. This was something that hadn't happened for so many months—ever since she had been ordered to leave her family home.

CHAPTER 15

The next morning, Lina woke up very early; in fact, it was too early. It was very quiet. The light in the windows was dim, but not the same as it looked late in the evening. Something filled it with formless twilight haze. She looked out of the window. Long, pale curtains of steady rain hung from the sky. In the curiously cold, watery-blue light in the courtyard of the garden she saw her workload—boxes of fruits she had to sort out: the best for the family and the worst for the farm animals. The wind was blowing through the gaps in the window making Lina feel as if she was inside a tube. It tried to blow away night's fog. Glancing upwards, Lina could see scraps of clear sky. She was hoping for the rain to die down. It was getting lighter, but it was still too quiet in the house and too early to leave her bed and start her day.

Lina knew the workload was already waiting for her. She would start with cleaning the windows, scrubbing the floors, and washing loads of dirty laundry that had accumulated in the wash room. She was overloaded with work, which left her with very little time to rest.

Her owner had a shop, so she had to clean and help to arrange items on the shelves. Alfred brought some new stock into the shop, and Lina helped to take the items into the storage area at the back of the shop. There were so many things, even biscuits and sweets. From time to time, she would sneak a handful of food and put it in her pocket for later. She treated herself without feeling guilty.

Today was a usual day for her. Whilst she was carrying out her daily duties in the shop, the door opened wide. A group of young women came inside. They were buying some groceries. Lina overheard Alfred asking them something. She stopped doing what she was doing to listen to their conversation.

"Where are you from?" he asked.

"From the Ukraine," one of the girls replied.

"I have a young woman from the Ukraine working here. She is here now, and she's from Zsytomir. Wait here. I'm going to call her to come and meet you." Alfred went to the back of the shop. "Lina, come here. I want to introduce you to someone."

After the introductions, Alfred left the girls to talk.

"Hi. *Dobriy den dorogenki*?" said Lina

"*Yak vam szivetsya?*" (How you finding living here?), asked one of the girls.

"I'm from the Zsytomir region," said Lina. "The village called Babrichi. I don't know if you've heard of it." Lina desperately wanted to speak in her native language.

"We are from the Kiev region," said one of the girls.

It was good to meet the people from the same country she was from. After a short conversation, it was time to return to her work duties.

Lina often saw these girls coming to the shop to get a few items. Lina exchanging a few words whilst working wasn't a crime in Alfred's books. He didn't mind as long as all the work was done. The girls told Lina how young men and women who were working in the factories were losing fingers due to no safety regulations. The safety hazards were everywhere due to lack of regulation. The young women were in the same situation as Lina, far from home. The regular visitors made purchases in accordance to a strict rationing system. It was very limited, and only items listed on the tokens were allowed to be purchased. Restrictions were everywhere—meal times, food intake, times to work, times to sleep, times to go out. These restrictions robbed the factory workers of their own free will, and they were exhausted from the long work shifts. They knew it was impossible to turn back even though some were trying to escape, but they didn't succeed. The ones who tried were shot on the spot. They chose to serve in an obedient way, driven by their desire to return to their motherland one day.

By now Lina had earned a good reputation for her hard work and good character. Alfred and his wife were impressed. In return, he didn't mind Lina talking to the girls from Ukraine. On her days off, Lina would go out to meet with them. Once a week she could meet her cousin Katya. It was the best time for both of them.

Today was Sunday, and Katya was due to visit her. Lina was waiting in anticipation. Suddenly there was a quiet knock on the door of her room. Lina ran to open the door. She let her cousin into her room and hugged her tightly. "Close your eyes, Katya. I have a surprise for you!" Her cousin did as she'd been told. "Now, open your eyes!"

In excitement, Katya's face lit up with a big smile. She couldn't believe it. Her cousin was holding some biscuits and sweets! Lina had not asked Alfred for these little treats. She had been taking them secretly from the shop without anyone's permission. She saw it as a reward for the work she did for the family. The girls would share the news and often would have a little cry together remembering all that had happened and not knowing what was going to happen. Both girls deeply missed their families. Right now, they were holding onto the good memories of the past and counting the days when they would meet loved ones again.

Life had treated Katya in a slightly different way. She was working for the owner of a watermill. He was very tight and was rude most of the time. He didn't feed her properly, and he was stressed all the time, forcing her to work endlessly. Katya cried during every visit with Lina, telling her cousin how bad he was.

"One day I was washing a window," Katya began, "and I accidently smashed it, so the sons of the windmill owner sent me to the German office where I had to wash cars all day. I was so hungry because they gave me only one meal a day. My hands were aching right up to the elbows with a constant, dull pain that was made worse because I had to wash the cars in ice-cold water. I think I have rheumatism. The punishment was for a few days." Lina hugged her cousin, and Katya continued her story. "The two days I was there were like two months. I worked like a horse without any breaks. I thought I would pass out! Finally, after watching me closely, the SS decided I should go back to the windmill. I wasn't glad to be back there. I was more looking forward to seeing you. I'm so glad to be back here," she cried to Lina.

Katya continued, "I was so scared to be there. I thought they were going to kill me with the overload of work. I didn't feel safe there, cousin." She was sobbing to Lina. She was shaking as she told of the horrors the SS had put her through. Her cheeks were tear stained, and her small frame trembled as she shared her ordeal. She looked so innocent.

They cuddled each other, crying and giving comfort, melting the sadness away. It felt as if the time was flying, and soon it was the time for Katya to go back to the windmill. She didn't want to be late because she might be punished again. Saying goodbye was always the hardest.

"See you soon. Promise you will be safe!" Lina spoke lovingly with encouragement.

"See you soon," Katya said. Lina walked outside with her cousin.

"I will stay here until I'm sure you've gone," Lina said in tenderly voice. She was so thankful to see her cousin whenever she could.

Lina stood by the gate until Katya disappeared around the street corner and she no longer could see her thin silhouette. She returned to her empty room, feeling unbearably tired. It was the weekend and she could relax. She thought about how strong she had become and how supporting her cousin was the right thing to do. It was good for them to be together sharing everything they could. It was easier for them. She clearly loved her cousin, and she needed to be strong for her. Lina knew Katya would do the same for her. With this thought, Lina fell asleep.

CHAPTER 16

One day, a new permission was issued that made a lot of young people relieved because it provided some degree of freedom. It stated that every working young person could leave to go to another city within a certain distance and spend some time there, but of course he or she must return by a certain time. So, on Sundays, many girls and boys would go to the city called Kerhaim. There were a lot of young workers there from the Ukraine and Russia.

During one of these visits, Lina's attention was caught by one boy who looked at her all the time. They liked each other from the first instance. The boy approached her and asked, "What is your name?"

"Lina," she replied quietly with a gentle smile.

"Where are you from?"

"From Zhitomir in the Ukraine," answered Lina. "And you?" she asked.

"I'm from the region near Vinnica."

"What is your name?" shyly asked Lina.

"My name is Peter. So, where do you work and how long have you been here?"

I've been working nearly a year for Alfred, a shop owner in Kemarkaem," replied Lina.

"Would you like to go for a walk with me?"

She didn't hesitate. He seemed a charming, interesting, and good-looking boy. After talking to this nice young man, she knew that he was working on the land, carrying out farming duties. His owner was very demanding, and the work was exhausting. The owner wasn't very generous. Peter was friendly and had a very good sense of humour. He did make her laugh a lot. She really liked him; he was very handsome and seemed to have a nice character and good manners.

Peter told her how young men were growing old and grey where he worked because of fear, malnutrition, and exhaustion. He had been working long hours without stopping. Somehow, he made her feel happy. He told her lots of jokes, and when she was with him it was easy to forget about her daily routines and difficult work schedule.

They started to meet up more frequently, and she continued to enjoy his company. Today they met up again. Before she noticed what he was doing, he took hold of her right hand, and he slipped something into it. "There is nothing to be afraid of," he said. "Just look at it for me." He spoke in a steady voice and with surprising composure. She opened her hand and saw a little golden coin. "It's yours," he said. "You might need it in times of difficulty, and if something happens to me, you know what to do with it."

He began telling Lina the horrific story of what happened to him when he was brought into the camp. Somehow, he felt he needed to share this story with her. "When we were brought to the hovels, I and some other lads decided to escape from there because we couldn't just sit and wait for the SS to take us into captivity. We did not want to just surrender and be taken so easily. We wanted to fight for our freedom. So we were determined to find and join the partisans. We wanted to be in the army and fight the bloody fascists in order to free people. So while the guards were changing their duties, we took advantage of the moment and ran away into the forest. When we were a fair distance from the hovels, we thought we could relax, so we started walking talking normally.

"On the other side of the road a boy was riding a bike. He was wearing ridiculous shorts and a scarf around his neck. He started shouting Nazi slogans and putting his hand up in the air. He was holding little knife as if threatening us. We were chatting, and he kept on giving us intimidating looks. Then he rode off. He must have felt obliged to report the total strangers he'd seen to the SS police.

Not long after this, we heard dogs barking. The barking and clatter of boots echoed all around us. The danger was never far away. The German police prowled the streets day and night, imposing curfew hours. Breaking the curfew could cost you your life. Apart from the German police, everybody watched everybody else, and a network of spies was everywhere, all reporting to the German authorities. It was very serious, and our lives were in danger. We knew we had to run. The SS were looking for us. They knew straightaway that we had escaped from the camp. The only solution was to run as fast as we could so they couldn't catch us. We ran. Our backs were breaking. Our legs were giving way. We were desperate not to drop for fear of them catching us. Suddenly we had to stop running because we'd ended up on top of a cliff. There was nothing left to do but to jump off the cliff. One of our lads—Fedor—was very fit. He said we should jump and bend into a curled position when we landed; otherwise, he said, we ran the risk of broken bones. There was no turning back, and jumping was the only chance to get to the other side and escape from the chase.

"Fedor went first and gave a sign that he was okay. The others, including myself, followed. Soon we were at the bottom of a cliff. Thank God everyone was okay. We decided to look for somewhere to hide. Fortunately, there was a hideaway in the rocks that looked very discreet, so we decided to stop there and wait until the night. We thought it would be easier to move in the dark.

"As soon as the darkness took over the earth, we decided to move quietly. The hunger and thirst had begun to take over our bodies. We ate some plants—leaves—and sipped some dew left on the foliage. It wasn't very nutritional, but it was the only way we could get some sort of food down our stomachs.

"We kept moving in the night and resting in the daytime for over three days. We were so thirsty, and there was nothing to drink. It's embarrassing, but a few of us had to drink urine. It had a sweet taste, and it was warm. I never thought I would have to do that. I could only sip few times. One day we reached a village. We tried to find a water well. After searching everywhere, we found a little old well hidden behind an old building. As soon as we found it, the lads began to drink pure, fresh water with such greed. They poured it all over themselves. It was as if we had found treasure. We all gulped gratefully. The excitement on our faces as we bathed and drank almost turned the clock back to the time of our childhood. We splashed water everywhere, not saving a single drop. It was a moment of joy. We hadn't had such good time for ages. Enjoying every minute of it was beyond any explanations.

"Then we decided to start moving again. No time could be wasted in our situation. Soon we were on our way heading out of the village. As we ran a bit further along, we heard the noises of cars and barking SS guard dogs. It was clear they were looking for us. Somebody must have seen us in the village."

Peter was telling his story in such detail, Lina could vividly see it. The captivating story left her with a half-open mouth as she listened to every word he was saying. "So what happened next?" She wanted to hear more and more.

"They captured us because there was nowhere for us to escape. There were fields on either side of where we were, and we couldn't hide anywhere. We hadn't eaten for days. We were exhausted and chilled to the bone. Our adventure had taken its toll. We began to feel weak. The SS caught us. We were shivering with shock and fear. The wind was biting through our thin jackets. Our breath appeared as vapour in the cold air. We were so many grey faces under a grey sky, which was covered by a layer of sick clouds.

"The SS took us into vehicles one by one and beat us so badly I still have some scars. My mouth was full of blood, and I could barely swallow. My entire body was in agonizing pain. I felt as if I was paralyzed by the pain from the beatings. They didn't care whether we lived or not. Clubs, boots, fists—the insane brutality was yet another test of our physical and mental stamina to see what we were made of. Some men who took the beatings did not get up. The actions of the SS men had everything to do with establishing superiority and humiliating the slave labour at their disposal. In the morning they took us to what seemed to be a camp. We were ordered to stay in line and walk behind the SS man. He walked us to the gallows where some dead bodies were still hanging on the ropes. There were some empty nooses." Peter's lips were trembling now as he spoke. "The officer looked at each one of us and pointed at me—no one else but me. He asked me to come forward. So I did. I could not feel the ground under my feet. I was completely terrified and felt nothing but cold inside. I thought, *This is it. My life is over.* I thought, *I am going to die.* I cried and prayed to God to save me. My legs were shaking, and my whole body was weak from lack of food and water. I could hardly stand up. But I knew I had to because he could have shot me on the spot.

"The order was to put us all in rows. We were standing behind the SS men, who had their fingers firmly pressed on the triggers of their guns. The mad barking of the dogs sent chills down my spine.

"All of us were thinking the same thing. *What do they want? Why are they behaving like this?* There was a rising panic in all of us, but nobody moved. Any erroneous movement could cost us our lives. To break out of the line meant a bullet. It was horrendous. We were fearful for our safety every minute. It was a thin line between life and death.

"Suddenly a window opened in one of the buildings, and a woman spoke to the SS man in an aggressive tone of voice. The SS man, with his huge nose up in the air—it reminded me of a cactus—listened to her with an annoying expression on his face. He kept nervously adjusting his glasses. The woman continued to shout in a psychotic tone. She seemed to be anxiously and persistently trying to stop him. She wasn't going to stop. I understood that the conversation was about me. My German isn't that good, but I could understand that she was saying something about her son and the war. She was very firm and would not leave the SS man alone until he ordered me to step back into the line. I knew that whatever this loud woman had said had made him change his mind in an instant. She had saved my life. That fact came to me straightaway. I will never forget her. I had never seen her or spoken to her before. If it hadn't been for her, my body would be dangling from one of those spaces amongst the others we saw on the ropes. I felt numb with shock. In a split second, my life could have ended. The realization was for real, Lina."

Peter breathed out and then filled his lungs again with air, as if he was afraid of running out of oxygen. He could still remember the horrors of everything that happened. He was sharing with her his ordeal—the severity of the danger that had nearly cost his life.

"So what happened after that, Peter?" She wanted him to breathe normally. She could see that telling his story troubled him.

"It felt like an intoxicating numbness penetrated my body," he said, "making me feel so shocked. After about fifteen minutes I started feeling better. The sudden appearance of the woman in the window had been some kind of miracle—her hysterical voice, her screams, her shouting. I believed she was fighting for my life, for my survival. To her I was a total stranger, but her reaction to the situation I was in made it seem that I was very dear to her. She was German, but she saved me from the SS death commando, who was also German. For a few moments, I was just so relieved and happy to be alive, I can't tell you how good it felt to be saved!" Peter put his hands together in a holy gesture.

"The SS are obsessed with things like this," he continued. "They like to play mindless games on real people. And if a few work slaves died, they couldn't care less. The skinny ones couldn't work hard, so physique was a factor too. Everybody who was stumbling about, shocked, dazed, and unable to move quickly enough was hit with sticks or bitten by the snarling dogs, their ivory teeth sinking through clothes and flesh.

"Still, after all of this, I was determined to escape again. Every day I waited for the right opportunity. But it did not come soon enough, and I wasn't prepared to take on a big risk. But then one day we were ordered to go inside the train wagons. No one knew where we were being taken. The punishments for escapees were severe—a slap around the face, a cane across the back, a bullet in the head. Once we were inside the wagon, we found that removing some wooden panels in the floor wouldn't be difficult, and it was a great opportunity for us to escape. There were three of us—Grisha, Constantine, and me. We decided now or never, the sooner the better. The risk was high, but we prepared to take the risk. We planned the escape carefully. We could not afford any errors, as even a small error could cost our precious lives. But if we stayed in the wagon we would surely die because of the daily beatings and torture. Focusing on pursuing the escape was a risky option, but we did think about the end result—freedom from SS wooden grave train wagons on wheels.

"Occasionally the train stopped at the stations, so we had to create an escape route before the next stop. Using our muscle strength as well as our thinking, we managed to pull all the nails out, but we kept the floorboards in place covered with straw, just in case the doors were opened and the SS men wanted to check on us. The escape route was decided, and it was just a matter of waiting for the next stop. The only thing that made it difficult was that the train would be stopping next in the daytime, so it made our escape more challenging. It was risky and difficult. Then, at night, we heard an aircraft, and the bombing started. Suddenly the train stopped, and there was a lot of destruction. It was the perfect moment, and we took the chance to escape. It was one of these moments—now or never. So one by one we squeezed through the escape route, staying underneath the train for a few moments, waiting for the right time to move again. It had to be done in an organized way. We had to move discretely in order not to attract much attention. Whilst the SS were busy running and shooting, it was a perfect moment for us to escape. We ran into the woods. I gritted my teeth as the tears of pain and shock ran down my face and dropped to the ground. The taste of salt penetrated my lips, but I had to keep on moving. We made a procession of escapees running towards the forest. After the initial panic and chaos at the platform, a sense of calm had descended near the train. We could only vaguely hear the shouting voices of the SS men in the distance.

"Then, suddenly, and somehow reluctantly, everything fell into place. We heard the snarled orders in guttural German: 'Raus! Raus! Get off the fucking train! Scheissjuden! Move it, Dreckjuden! Shit you! Dirt you! Keep it quiet, you scum! Get over here, line up!' The occupants of the trucks came from everywhere and it occurred to me that they couldn't understand what was being said to them. None of them could have missed the essential message however, because it couldn't have been clearer.

"We were running. It was hard for everyone to keep up, but adrenaline kicked in and kept us on our feet, running for our lives. We didn't go into any built-up areas this time. We didn't want anyone to report us again. The young people had been trained to report any foreigners or strangers to the SS. The SS would know they were escapees and would send them directly to the gas chambers.

"We had trouble finding water and food. We ate anything that was edible or that we could get our hands on, and we drank rain water from the leaves and puddles. It was too risky to enter the built-up areas, as we would be inviting death. The dogs would smell us straightaway, and the prowling SS men were alert to pick up any sound. Thinking cautiously was our plan.

"Grisha started to show the signs of illness. He was getting weaker and weaker and finally could not move. He was hallucinating and losing conciseness. We did everything we could, but he had been weaker than the others at the start, and in the end the poor lad gave his soul to the Lord. We buried one of our lads and carried on. There was no time for stopping. We walked many kilometres, covering large distances. Crossing a field we found a dead man in uniform. His body was so swollen, we knew he had been dead for some time. Constantine noticed a watch on his wrist. It was stuck in the swelled-up skin. It was difficult to remove the watch, but he took it off and put it in his pocket. He thought it might come handy to exchange for some food.

"Soon we reached Pisikgaem. We saw a church and we decided to go inside the place of worship. It was the only safe place where we could hide. Very carefully we moved towards the church. We cautiously stepped inside the magnificent place of peace. I experienced a feeling of safety the moment I entered the holy place. Something made us fall onto our knees. We prayed for hours. The tears were rolling down our haunted grey faces. We were trembling with panic as we looked anxiously at the icons of saints across from where we were kneeling. Our eyes then fixed on the cross. It was a moment of healing and restoration that mended us the way no other experience could. It was a pleasant, warm feeling, that sense of calmness, peace, and safety as we were shielded from the harsh realities of Nazi captivity. It was a place where we could hide from the hungry, never-satisfied SS officers.

"Suddenly, the sound of footsteps outside made us want to hide. We hid in the vicar's confession booth and watched through the little holes in the fabric. The place of confession had been the nearest place to hide. The vicar was old. He reminded me of my grandfather. The lines on his face indicated the goodness of character. His warm, friendly facial expression indicated that I could trust the old man. I just felt that I had to come out from the hiding and confess. Something prompted me to do it.

"I fell on my knees and spoke in broken German. The tears poured out of my eyes. I could not stop the flow. I begged for safety. I begged for protection. The old man showed me a sign by placing his hand on my shoulder. It gave me a sense of assurance. He was a godly man. I could sense incredible peace, and then in an instant I blacked out.

"The next thing I remember is waking up in a very dark room. Constantine was sitting beside me. 'Huh, glad you have come around,' he said. 'Drink it.' He handed me a warm cup of some medicine. It smelled like an infusion of herbs. 'What happened? Where are we?' I asked, certainly in a state of confusion. I had had to find out what had happened to me and how I could possibly have blacked out.

"'We are in the church, and the vicar said we can stay here,' Constantine told me with a stern look. After all we only had each other and had to look after one another. 'He said we have shelter here while you are sick, but as soon as you are better, we must leave. Just drink this and eat something, and we are going to have to think of the next move.'

"The vicar came to see us. He gave us some clothes. He told us we had to blend in with the rest of the people as soon as I was fit enough. He said that, in the town, there were many youngsters from the Ukraine working in the local factory, the fields, and the farms. We had to find a way to blend in without drawing too much attention. We would look like the other workers, and no one would notice.

"So, one day we left our hideaway and walked into the town on the day when all the young workers had their time off. Keeping our heads down in order not to draw too much attention or show any signs out of the ordinary, we carried on walking even though we were very tense. Constantine and I began to relax a bit more when we saw all our lads and girls. We talked carefully to a couple of guys who were standing in the street and asked where they had come from. And it turned out that they were from Vinnitsa. We shook hands and asked them to help us with work. They asked us where and for whom we were working. We told them we worked in the local church but there was no more work left for us, so the vicar didn't mind if we worked for someone else. It was the best answer we could come up with.

"We could not tell anyone about our escape. We still had to be very cautious, discreet, and careful even with the people from the Ukraine. It was evident everywhere that much work was needed on the farm; they could do with a couple of more lads. They took us into the farmyard and walked directly to the farmer. The farmer was a fat, red-faced man with ginger hair. He looked up moved his hat and said, '*Was is dos?*' 'I've got more lads for you,' one of the guys said. The farmer looked at us from top to bottom for a moment as if deciding whether to take us on board or not. After few moments, he made his mind up. '*Goot! Goot!*' It did not bother him for whom we worked and where. He didn't even ask any questions. He ordered the lads who had brought us to show us all the work, and he carried on with what he was doing. We had caught him at the right time. Constantine and I just exchanged glances, saying nothing. Inside we felt great relief.

"Then the work started. We were covered in dust and dirt and clinging sticky mud. We looked as if we needed a good wash, but the days of hard labour left us with no chance to have a wash. The lads looked after each other, giving a hand if someone needed help. One lad from Russia couldn't handle the work. He collapsed in the field.

Later on, they brought him on the stretcher, his arms and legs dangling disturbingly. He didn't make it and was put to rest the next day. We all attended the funeral. It was a very quick ceremony. Taking our hats, off we paid our last respects."

"So what about the vicar?" Lina asked. "Did you see him again?"

"Oh, yes. We went and told him we were working. He was pleased that we had blended in. We found out that the old chap passed away very shortly after our visit. May he rest in peace." Peter crossed himself, demonstrating his faith in the highest of all. "We will always remember him for the kindness and protection he gave us," he said with evident sadness. "Without him, who knows what would have happened to us?"

Peter hugged Lina with gentleness. "It feels so good to be here with you. We can share everything," he said. He felt as if he didn't want to let go. Being with her melted away all the anxieties and tension from the horrific events he had told her about. Peter kissed her soft lips and then her cheeks so gently, sending goose bumps all over her body.

She felt very warm and comfortable being in his arms. She felt that she was in safe hands. She and Peter had shared a special connection from the moment they met and began telling each other their stories. Both were far from home and had become captives of the harsh realities of the Second World War. They were young and inexperienced. They had been taken by force. It had not been their choice. Even their parents could only warn them not to run, or they would suffer severe penalties that could cost them lives. The SS were obsessed with watching everyone very closely, monitoring every move, every word, and every action. Anything out of the ordinary meant a bullet straight to the head— a weight gain of nine grams. The farmers had strict orders to report any suspicious behaviour. Everyone knew about it. Somehow the workers stuck together, and they all looked after one another.

CHAPTER 17

Lina and Peter saw each other as often as they could. The farm owner for whom Peter worked wouldn't let him out very often. He overloaded the boy with many duties. He restricted his workers with curfew hours, and there could be severe punishments if the rules were broken.

As soon as the young couple had time, they saw each other again. One day, Peter told Lina about the previous week's bombing in Pisikgaem. "I was sleeping and then suddenly the airplanes came and started dropping bombs. The farm owners ran and hid in the bunker. They shouted for me to do the same. So I waited until they left the house so I could grab a crust of bread to eat. I was starving, and that was the only chance to grab something to eat."

He continued telling her what happened. "I ran into the kitchen and was grabbing some bread when I heard the owner calling for me. I put the bread back where I got it from. I wanted to eat it, but if the owner saw me taking it without permission, I would be a dead man there and then." The food supply was becoming increasingly hit and miss as the war began to pinch the Reich, and even those who were considered to be "privileged" workers felt it. "I was so hungry," Peter said, "and the starvation was killing me, but I couldn't risk it. I quickly ran into the bunker to hide from the air raid."

Lina was captured by Peter's stories. She was a good listener. When he told her all about the air raid she felt like pouring her heart out to him. "Oh, Peter, for how long are we going to suffer? When will there be the end to it all?" Lina was full of emotion. She was deeply disturbed and troubled. She was sensitive to the way she felt about Peter. It troubled her to know how he was being treated. When he was hurting, she sensed his pain. Their relationship was pure. They truly cared for one another, but they both knew that, at this stage, it was more than just a friendship. "I'm very worried about you," she said in a quiet tone of her voice.

"Darling, don't worry. I am going to be fine," Peter replied with a smile. He was always smiling. It was exactly what had attracted her to him. It was fascinating the way he was always smiling. Somehow it took away the worries she had, melting away all the anxieties she had had to deal with, burdened as she was with a heavy workload. "The farmer must feed me no matter what because he relies on his strong workforce. He knows that men become weakened by slave labour and starvation. It is a lethal combination. If the workers lose their strength, then the farmer will have only weak men who will not be equipped to do all the hard work. He has two daughters. One of the daughters likes me, and she seems nice. Both, however, are so lazy. They don't want to do anything. The other hates me and is always shouting orders overloading me with the heaviest work. One day the pump had to be operated to pump the shit out from the toilet. So I pretended I was too weak to unscrew the handle. I asked the one who hates me to help me out. Anyway, with her usual unpleasant and sharp look full of anger, she unscrewed the top too fast—it must be done very slowly, carefully. The shit covered her from top to bottom! I had to hold myself from bursting out laughing seeing her unpleasant face completely covered in shit. She quickly ran inside the house screaming. I didn't see her for weeks. It was hilarious. Her sister told me that the girl had to soak herself in the bath to get rid of the smell and get over the shock."

Peter's story made Lina laugh. With their sense of humour they supported each other to relieve the hefty burdens of war that had landed on their young shoulders.

As Lina worked in Alfred's store, she often took a few "luxury" items to give to Peter—biscuits, cubes of sugar, anything he might like.

"My sweetheart," he said, squeezing her shoulders tenderly, showing his appreciation for caring, "I am going to be fine, just fine." She, in turn, reassured him about her well-being.

Peter knew that she was concerned about him, and he liked the fact that she wasn't afraid to risk taking food from the shop for him. Her toughness of character, which came from her father, was very well evidenced in her actions and character. "You have to look after yourself," she reminded him repeatedly.

After a short pause, she began telling him about the changes that had taken place since she had seen him last. "I'm working for two families now. I was sent to work for another family of farmers. They are relatives of Alfred and Anna. The farmers have lots of land, and can you believe it they use oxen for all the work in the field? They have no horses like we do. I clean the house, wash the clothes, and help with picking some of the crops. The farmer's wife is called Hana, and she's Alfred's wife's sister. She lets me to do almost everything. I can eat anything I want and even cook for myself. Her husband looks after the land and spends most of his time in the field. He works hard for his family. His dedication is proven by the amount of land they have, their huge house, an orchard with apple trees. Not a single apple is wasted. No apples rot underneath the tree, making it look untidy. Everything is picked up. The property is immaculate almost to perfection. It is a very tidy family-run farm.

"I'm very busy doing the endless chores in Alfred's shop," she continued, "plus looking after two houses, helping Anna and Hanna. It is very hard for me—twice as hard as before to carry out all the work. But I look forward to seeing you and Katya. Seeing you both keeps me going. And knowing that one day we can return to the Ukraine. You cheer me up so much!" She gently pinched his cheeks touching the end of his nose with hers. "Here. I brought some treats for you. Eat. Don't be shy. It's all for you. Eat and take this with you. Sometimes I can take some items, not that I am allowed. I do it. I know they don't mind. Alfred seems to notice my hard work. He respects me for all the work I do. His wife could not cope with it all. She is a different kind of character, very firm with high expectations, lots of orders, endless demands. She doesn't like physical work. She only likes giving orders to other people, including her husband. She keeps me running all day. Some days I am rushed off my feet! Its God's truth, I am telling you." Lina was smiling with some intensity.

The more time they spent together, the closer they became. With the war going on, it wasn't for them to decide when they could meet again. The curfew hours, demanding working conditions, and endless jobs took them to the unknown. They didn't even know whether they were going to see each other next, but they believed in better times. Their lives were so unpredictable and full of uncertainties. Despite all of this, they were happy when they had the chance to get together to walk and share stories about the past and present, and only dream about the future. They took each day as it started, with an incredible obedience. They couldn't stay together; they would be separated by the cruel demands of the war.

PART 5

LIBERATION

CHAPTER 18

There was talk that the war was going to end, but still there were no real signs of any changes. No one knew how close freedom was until one day when the American army rode victoriously into the village of Kemarkaem. The throb of the truck engines made the walls shake, alerting everyone. It was quite an unexpected indication that the times were about to change. This was when many youngsters were freed from their long, hard, compulsory work. Most were given freedom to return to their home countries or to travel anywhere else in the world to start a new life. The United States of America opened up its borders for those who didn't want return to their countries of origin. It was a time to make big decisions, a time that would change their young lives again.

It was the time of new beginnings and the end of enforced hard labour. The heavy chains of fascism that had forced devastating separations, obligatory controlling rules, and slavery had been broken. For most, it was liberation from the life they had known day in and day out for years.

This was the first time Lina saw a black man, an American soldier. He stood in front of her in a green uniform with lots of equipment, he was fully prepared for the military mission. On his left side she noticed a grenade attached to his web gear. He was holding a gun, and his finger was firmly pressed on the trigger in case of unforeseen emergencies. A green scarf was carefully wrapped around his neck to keep him warm. He wore a helmet with a strap around his chin to hold it in place. Lina felt frightened when she saw this man. She didn't know what he was about to do. The soldier released his finger from the trigger and took a rucksack off his shoulders. He opened it up and took out some tins, showing Lina that it was food for her. He was trying to talk to her. She was trying to understand his gestures. She understood that the soldier was asking where she came from. So with hand gestures, she tried to show that she came from a far, faraway, saying, "Rusland. Rusland." She repeated it so he could understand her.

The soldier's smile lit up his face and exposed perfectly white teeth. He had a generous face, and the laughter lines made him look like a friendly wide-framed giant standing in front of them giving away his food to feed the undernourished fragile bodies. He continued to give out food from his huge rucksack, making sure that everyone got something before he left in a hurry. The fully equipped American soldiers came and freed all the youngsters from their long-lasting captivity.

Getting their freedom back was like getting their lives back. It was the beginning of a new fresh start. Each one could go his or her own way. Some were heading for Europe, some to England, and others to the United States of America, but most of them were heading to their motherlands, back to their families. They had survived servitude because they knew that, after they served, their freedom would be established. It was the day of their victory, their liberation, and their independence.

Lina was near hysteria at the prospect of liberation in 1945. Her heart was filled with happiness and unspeakable joy. Everywhere was hectic. She could hear laughter, dancing, and the lively tunes of the accordion. It was the happiest moment for most of the young people. That day the Russian General Ivan Ronev took Krakow and Jarnov, the news reported, and Marshal Georgi Zhukov liberated Lodz. Full of excitement and new hope, Lina walked through the crowds searching for Peter. He was nowhere to be seen. *Where is he?* She thought. *Where can he be?* She was speaking to herself, desperate to find him. Suddenly, someone grabbed her by the hand. It was her Peter! She hugged him with such enormous strength.

"Congratulations, my sweetheart!" he said squeezing her gently. "I'm so glad to see you here!"

"I was looking for you everywhere," she said. "I couldn't find you for hours."

"It's very important! I've got something to tell you." His face turned very serious now. She believed he was about to say something of great importance, and it alerted her to prepare for what he was about to say. "Lina, I want to ask you to go with me to America."

His eyes were piercing as he looked at her, and she found it very difficult to say no. "What?" she asked, pulling away from him, too frightened to give an insincere answer.

"We could go with the American army now and start our lives afresh, just two of us, you and me.

The American soldiers are here to free us. They are giving us food and even ammunition for our own protection."

"Why would they do it?" she asked with curiosity.

"Their intentions are good." He was smiling. "Don't see what they are doing for us already? Look, now is the right moment. Let's do it together, sweetheart. Come." He took her hands and squeezed them tenderly. "Come on, my love. We can do it together!" He was so confident, and there was not a sign of hesitation on his face—none at all. He was sure about leaving with her right at that moment.

"Ah, no, Peter!" she said, turning her head from side to side. "The last thing I want to do right now is to go to America." She spoke with the surprising firmness and not a single bit of doubt in her voice. "I must return home. I have not seen my family for years, and I'm dying to see them. I cannot go to the America now. Do you understand me? She was trying to convey exactly how she was feeling. She didn't want to give him any false hopes. "Of course, if you want to go you can go and do what your heart tells you to do." She was trembling inside but didn't want to show too much of her nervousness. She was thinking to herself, *If he decides to leave me and go to another country, I will never see him again, and that would break my heart.* Either way they would have to be apart. It was not easy to accept their circumstances, but they had been left with limited choices. Still trembling, somehow she was prepared for anything, even if it meant they were going their own ways.

Peter was thinking. He was trying to work things out—what course of action should he take. Taking her hands into his he said, "Oh, Lina, Lina. No, I can't go without you. We're destined to be together. Do you understand me? You're my love." He hugged her tenderly.

She felt his soft warm lips. They shared long passionate kiss. "Then don't go to America, and we can stay together," Lina whispered quietly.

"I'm not going without you. Full stop," he replied. "Do you understand? Let's go." He took her hand firmly. He wanted to find some of his mates and find out what they were planning to do. Peter spotted several lads he knew. With all their belongings, they were queuing up. "What are you queuing for, lads?" asked Peter.

"Can you see the registration point on the other side?" A lad wearing a cap pointed with his finger. "Can you see, just over there?"

"To register for what?" Peter asked curiously.

"To return home," the lad replied. "The government has ordered us to join the Red Army, and the sooner the better. It's obligatory in order to repatriate."

"Here it comes again," another of Peter's friends said. "Out of one captivity into the next one."

If we must serve, I would rather serve in the Red Army," Lina exclaimed without any hesitation.

There was so much to think about, and there were so many decisions to make. There were crowds of queuing people—those who had decided to return home and those who were beginning their journey to Europe or to America. The area was a mass of chaos and excitement. Tears and laughter filled the streets with a noisy hum. It was a time of freedom, but also one of those moments in life when directions become twisted. Nobody knew where each of them was going to end up.

Now that it was clear what they were doing, Lina had to return to Alfred's and Anna's house to collect all her belongings.

Peter joined the queue to register. Going to Europe or to America was out of the question now. He made his mind up and was very firm about returning to the Ukraine.

"Listen Peter," said Lina. "I have to go and get my things."

"I'm coming with you," he insisted.

"No, no. I will be fine. Don't you worry."

Just before she was set to leave, Peter took her hand. "Take care," he said, kissing her hand. Her hands were soft and incredibly feminine. They smelled deliciously sweet. "I will be waiting for you here, my sweet. Just get your things and mind the way you go," he said before letting go of her hand.

In just a very short time Lina was at Alfred's house. Anna, his wife, looked very grumpy. It seemed that she had been waiting for Lina. Trying not to look at Anna directly, almost ignoring her, Lina quickly ran upstairs and started packing all her clothes. She just about threw everything into her suitcases, not paying any attention to the neatness of her items. She closed the suitcases quickly and was on her way to get out of the house as soon as she could.

Anna was fuming. She snatched both suitcases from Lina, opened them, and pulled some of the dresses she had given her to wear when she was working. Her face turned red from anger and frustration. Her entire body was nervously trembling; her hands and legs were shaking. She looked like a nervous wreck. The sudden changes in their village had made everyone worry. Who were these people in uniform, and what would they do to the villagers now? How would things change for each one of them in the village? The thought of payback had set their pulses racing. No one could predict the outcome. No one could have been prepared for whatever was going to happen next.

Lina stood innocently in the corner. She didn't even attempt to resist to Anna's sudden rage. Anna was a woman of a strong character. Even her husband would not attempt to touch her when she was in such a state of fury. She was dangerous and could destroy anyone who come near her. She possessed great shooting skills and had a gun. No one would mess with her. All the villagers knew—even Lina knew—to keep quiet during Anna's sudden bursts of rage. Not a single thief would attempt to step on their land. They were terrified of the woman.

When Anna finished retrieving the dresses and throwing the suitcases on the floor, Lina obediently picked everything up and walked out of the house, firmly closing the front door behind her. She was relieved to be out of the house and on her way to new endeavours.

Alfred was nowhere to be seen, but he was the only person she wanted to say goodbye to. He had shown her compassion because he had seen her as young woman who worked well despite being far from home. She had served his family day in and day out without complaint.

Closing that door behind her felt so good! Lina walked fast, her legs picking up speed, and she didn't even realize that she was running. She was running, leaving her past behind. She was firmly on her way to the freedom. Tears of happiness were streaming down her face, and she felt an unusual lightness as if she was almost flying. The wind was blowing in her face making her hair escape from the scarf that ordinarily kept it neatly in place. She was thinking about Peter and trying to get to him as fast as she possibly could.

Finally, she saw him. He was waiting for her exactly where she had left him. When she stood next to him, something made her think he was deeply concerned. She could see the lines on his forehead, and his strained jaws were moving from side to side. These facial features were signs of tension, and she knew something was wrong. "What's the matter?" she asked

"I've just been told that all of us must join the Red Army, and you and I are going to be apart, my sweetheart. It's a new order from the government," he said abruptly.

"For how long?" she asked in disbelief, raising her eyebrows.

"Don't know. Not sure. No one knows." He hesitated to say anything else because he didn't want to upset his sweet girlfriend. "I was told that we would have to serve until the war is officially over."

Lina wasn't smiling. The unexpected news gad wiped the smile out of her face. She looked into his eyes for reassurance. "But the army—the soldiers—are here to set us free!"

"Yes, if we want to go to America," Peter clarified. "The soviet government wants those of us who want to return to the Ukraine to serve in the army until the fascists are completely wiped out."

"Everything is not how I saw it," she said. "Never mind. We will have to go with it." She adjusted her thinking. In a way it made sense. She could focus on what was to happen next. "Someday, somehow, it might be clearer," Lina said with a facial expression that indicated her hopes.

"Let's hurry and register. We must do it now, my sweetheart," Peter said, gently pulling her closer to him. The registration and classification point was several meters away from where they were standing in the queue. He lifted their suitcases, and they both moved up in the queue closer to the registry point.

The humming noises coming from the people talking in the queue were so surreal that Peter couldn't believe what was happening to him. His life was changing once again in a way that no one would could have predicted. He felt a sudden, sharp pain inside, something he had never felt before he met Lina. He never would have thought it would be so very difficult to say goodbye. So many thoughts were going through his mind. He thought it was unfair for them to be liberated from one captivity and have to face yet another strict government obligation.

Lina was crying. She didn't doubt that he loved her because now he made it very clear. Holding her gently, he was wiping the tears that were running down her cute cheeks. She looked so innocent, and he didn't want to let her go—not now, not ever. He knew he loved her, and it was enough to know she was the one for him. He knew that there and then without any second thoughts. "Listen, we could go to America, England, Belgium—anywhere, if you wanted to. Just say so and we can. That way we could definitely stay together, my love." He was trying to think of any way they could be together rather than apart.

The thoughts were rushing through his mind. Neither of them wanted to be accused or penalized. No one there understood the new situation or fully grasped its significance. The Nazis had used slave labour as one of the many propaganda tactics to enhance the German economy. The enemy of the state sounded petrifying. Fear gripped Peter and Lina, clouding their thoughts, but they had to make the choice. And the choice they made would take them on additional life journeys as they served so they could be free.

For Lina, the thought of not being able to see her parents was unbearable. She had been taken from them in the middle of the night and escorted away by the SS. Packed tightly with other captives, like cattle in the black crow, she had been taken far away from home. Not seeing her parents would break her heart. She had been thinking about them all the time, and now she finally thought she would have a chance to see them, even if it meant working for the Red Army before returning home. She was thinking now and, after a short pause, she replied, "No, Peter, I can't. And it's only because I've promised to return as soon as I can. We are facing this separation, but it cannot last forever. The war will eventually come to an end." With a positive mind, she looked at him with much love, wiping the tears that ran down her face. "If we can get through this, we will be able to get through anything that life gives us." She smiled at him with much reassurance.

"Yes, my darling. We can, and we will," he replied firmly. "If we are meant to be together, we will be together."

Lina felt so comfortable in his arms. She didn't want to let him go. She felt butterflies in her midsection just standing close to him. When he touched her, she had the shivers, and their kissing was very passionate and warm. He had become so very dear to her. She experienced an unexplained lightness every time they saw each other. She knew she was in love. It was as if something had taken hold of her and given her warmth and comfort to her core. It was real, and she felt it was mutual. She didn't have any doubts—none at all. She just wanted to love the man, open her heart to him, and shard every moment they spent together. They had each other, and that was all they needed to survive.

The difficulties they experienced now seemed like everyday living. For years, as captives, they had known nothing but never-ending hard labour and curfews. Both had endured the toughest slavery far from their motherland. Their faith, determination, strength of character, and strong desire to live had kept them going. Their togetherness during the most difficult atrocities and an incredible loyalty and devotion helped them both get through the times of the most destructive and eliminative war in the human history.

It wasn't the end of their journey, however. They were now faced with separation as they were under government orders to wipe out the fascist rulers of Germany. It was one more obligation to fulfil, another order to carry out, and another way to serve.

PART 6

SEPARATION

CHAPTER 19

My Nan was trying to catch her breath after telling me, very slowly, about the moments of their liberation. She was persistent in conveying her story even though she often had to pause to think as she shared glimpses of the moments so distant in her past. Her wrinkly hands were trembling as she spoke. Her brightly coloured scarf, which covered her grey hair, was tied in a neat knot under her chin. She looked old. The years had taken their toll. Her facial features, however, showed evidence of her godly beauty. She looked bright and sweet, and she told me her story in the sweetest tone of the dearest voice. Her voice was comforting, like a lullaby to my ears.

I wrote very quickly, jotting words down because everything in her story was important. I could also remember some parts of the story because her husband—my granddad— had told it to us many times. She continued talking, and I continued writing, trying not to interrupt. When she would stop to try to recollect moments, I would also stop until she wanted to continue again. Our conversation was a revelation about the reality of her story and my granddad's story. Their youth had been stolen by years in captivity. The story did not just captivate me; it also instilled in me the desire to share it with the world. I hope it draws enough attention for people to know how devastating war scars can be and how miserably humanity can fail by initiating war, making costly mistakes over and over again. Historians today can't decide just how mad Adolf Hitler was. Some of them point to his narcissism, his arrogance, his obsessive racism, and his increasing megalomania.

The Second World War, they say, was Hitler's War. You don't have to look any further. *(Sam Pivnik, Survivor, Peter Conradi Sunday Times, HODDER, 2012)*

My Nan could remember fragments of the day when tank forces of the Red Army cut open the defence of the enemy and consistently moved forward, leaving behind the approaching infantry forces. The tanks moved forward without stopping, entering cities and towns, destroying enemy equipment and manpower. The Red Army moved through the land of fascist beasts, getting closer and closer to achieving a great goal that took years of passing through bloodstained roads. From the sky, pilots guided tanks through unfamiliar surroundings as they targeted more and more of fascist nests, which eventually led to the capitulation of Nazi Germany in the first days of April 1945.

Lina was transported to work for the Red Army. She had numerous duties—mending clothes, washing, and cooking. Her days were filled with lots of work leaving her with a very little time to think about anything else. She was a favourite of the whole squadron. She had the reputation of a brave young woman. She worked hard to get her chores done. When Lina joined the squadron, she tactfully but firmly showed no interest in all attempts of the soldiers to date her. She just became everybody's friend. In the evening, she would sit down at a table in the middle of the room with a pile of shirts to mend and sew missing buttons on.

"You have to sleep," one of the soldiers said.

"I will be on duty tonight, and at the same time I can mend the boys' clothes." Lina smiled at him. "They have to fly tomorrow. I can sleep during the day."

It was silent. Everyone was asleep. Lina was sewing and quietly singing a famous song called "Zemlyanka". She could hear the adjutant's watch ticking. Then, suddenly, the door squeaked, and she heard some noises. A gunshot. She flew to the door, but she could not open it. She tried to push it, completely forgetting that she should be pulling instead. She heard numerous screams and gunshots. The sound of machine guns. The adjutant jumped out of his bed as the door flew open. Everybody began to run out of the bunkers. Lina was petrified. In the doorway, dimly lit by the flame of her kerosene lamp, she saw three fascists with rifles in their hands, looking at her with wide-open eyes. Fortunately, one of Red Army soldiers woke up. Taking up his gun, he fired at the uninvited intruders, making them run out of the bunker.

When Lina ran outside, she saw that their worst fears had come true. A group of SS men was trying to take over the airfield. The firing of their huge machine guns drowned out the small gun shots of the Red Army. This was a real battle ground.

There was only one thought in the head of each pilots: *How do we get through the shooting to the bloody airplanes?*

Lina started to move carefully. She heard someone calling her name. "Stop!" Someone's voice insisted. But she pretended she didn't hear. "Stop! They will shoot you!" the same voice said. She didn't recognize the voice of the person who was telling her to stop. She didn't even understand why she should stop. *Should I stop moving or stop listening to others?* she wondered.

Lina decided to hide quietly in the dark. After a few minutes, a rocket was launched that lit up the entire surrounding area. She couldn't see any SS men in the field. They were hiding in the forest and were firing from there now. She saw the silhouettes of Red Army soldiers shooting towards the forest. The bloody fascists stopped shooting for a moment.

Much better, you bastards, Lina thought, hesitating to convey her thoughts out loud.

Everything went silent again. That was worrying. Everyone strained to see and hear what was going on. Lina worried that, in the darkness, the fascists could get to the Red Army's airplanes and take over. *We should carry on keeping them at distance with the machine guns*, she thought. *Then again, we can't waste our ammunition because, in the morning, the boys are due to fly.* The Red forces decided to carry on launching the rockets. The artillery division was at work keeping the fascists beasts out of the way. Their precise order was not to let them close to the airplanes during the night.

Very slowly came the sunrise. There was no sign of the enemy. *They must be very far from the airfield,* Lina thought. Everyone in the division was making sure that everything was under control. Fortunately, there had been no lives lost, and all the vehicles were unharmed, except for a few bullet holes from the night's shooting. To everybody's relief, they had been very lucky; it could have been much worse. The commander was reprimanding the shooting team for not spotting the fascists intruders. "The entire squadron could have been wiped out by those five intruders!" he shouted at the lined-up boys, who stood with the heads down listening with obedience, without uttering a word in their defence. Military discipline did not allow disagreeing with the commander's instructions. That ruling came from above, and everyone had to respect it so that clear instructions could be followed.

It the morning, the birds were singing. Spring was in full blossom. Lina wanted to go into the forest, lie down where no one could see her, and forget about everything. She wanted to think about her Peter, the only man she had ever loved. She wanted to forget the night's distraction and loud shooting and the fear of losing the squadron. Her eyelids were getting heavier from the night's attack, and she wanted to escape somewhere where no one could find her.

Radio transmissions informed them that their enemies were running away now. The SS, the Wehrmacht, the entire Reich, were on the run, and the end of the war was in sight. The SS still were trying to create a false illusion by continuing their attacks, as if to prove they still had the power of life and death over their enemies, and as if they would decide how, where, why, and when they died.

Lina's thoughts were interrupted by a loud voice from one of the mechanics. "Commander, the SS are over there!" He pointed to the side of the shed, approximately three hundred metres from the airplanes.

"What?" asked the commander.

"I went to the shed … well, behind it, commander—"

"You aren't making any sense, fella!" the commander interrupted. "It's all because you are just frightened, and you think you saw the enemy. It happens."

"I swear!" The mechanic was totally serious, and his response prompted the commander to send two snipers in the direction of the shed to check everything out. After several minutes, they brought the fascists out. Dirty, unshaven, and wearing ripped uniforms, they looked like total failures.

The commander's hand moved to his gun. "I want to shoot these bloody bastards so they can rot away right here! Each one of them gets it now, for each of the friends I lost. For the loss of blood and for starting this horrid war. For intruding onto our land, killing our men, women, and children!"

Lina saw fear in the eyes of the fascists. They knew what commander wanted to do right that minute.

"To hell with you!" the commander said with a sigh. He put the gun back into his holster. "I'll let you live," he said with the stern look on his face.

Lieutenant Kopotkov came along just then. He could speak German. The big ginger man in the SS uniform was trying to speak Russian. From the conversation everybody heard the phrase: "Hitler kaput! Hitler kaput!"

Korotkov interpreted that the fascists had ended up on the airfield by chance. Their group of five were what was left from several divisions that had been broken down by the Red Army. They were hiding behind the building ready to surrender.

Some of the lads from the Red Army squadron gave the fascists cigarettes. They smoked greedily. It was clear that they were still fearful of the commander's gun. Suddenly, one of the captives put his hand into his pocket and took out his diary. He retrieved a photo. Lina could see the soldier in the photo dressed in ordinary clothes. Next to him was a woman with two children. "Frau, kinder," he said, smiling.

Wife, she understood, and his two children. She looked at the happy family.

The captive was trying to say something. Korotkov interpreted. It seemed that the soldier lived not very far away. He was a chauffeur, and he had no intention of harming anyone. He had been in the army for only three months, clarified Korotkov.

"So what to do with the prisoners?" said the commander. "We must take them to headquarters for questioning."

"It's impossible to do it now, sir!" said one of the Red Army soldiers. "What if we just let them go?"

"It's risky," said the commander. "They could bring others and put us through a meat grinder. But something tells me they are not going to do that. That is not what is on their minds right now." The commander turned to Korotkov. "Tell them we are not going to fight unarmed prisoners."

Korotkov tried his best to interpret as clearly as he could.

The captives nodded their heads, saying, "Fershtein. Fershtein" (understand). They put their hands together as if thanking for the commander's decision.

Suddenly Lina spoke out. "Commander, why don't we feed them? Look at them. Their stomachs are stuck to their backs. They are people too."

"You are ready to pity even your enemies?" questioned the commander. "These snide bastards who took you from you family, made you suffer from starvation, took you away from all the good things? They could have shot you a few hours ago!"

"I don't want to harm even the worst enemy," said Lina. "Look at them, sitting right now on the floor with their hands up, surrendered."

"Right," the commander replied. "It's good that you don't make all the decisions, Lina," he cut short. "We can feed them, but with what? Where do we get the food? Are you be prepared to give up your own allowance, Lina?"

"Yes," Lina replied.

"And me," said another soldier.

"Me too," others replied.

"All right, take my food allowance too!" said the commander, giving in to the decision of the majority.

The prisoners, accompanied by two infantrymen, went into one of the bunkers. Lina followed them. Taking some food out, Lina put it on the table. The prisoners ate hungrily without saying anything to anyone. They were then given cigarettes and were ordered to sit in the corner of the bunker. They complied, smoking with greed as they looked from side to side.

The commander walked into the bunker. "Well, what shall we do with the prisoners?" he asked.

Korotkov told the prisoners that the war was coming to an end and everybody, sooner or later, had to surrender his weapons. The Germans nodded their heads. One German, the huge ginger man, tried to tell Korotkov that they had already thrown down their weapons.

"Go to your people," Korotkov told them loudly. "Understand? Tell them to surrender their weapons. Is that clear?" He waited for an answer.

The SS men talked angrily amongst themselves, pointing in the direction of the forest.

"What are they talking about?" Lina asked Kopotev.

"They are saying that in the forest are many more soldiers. They are fearful to come out because they are afraid of the Red Army.

Korotkov listened to the commander and said, "Okay. The commander says, let them go and tell their soldiers how the Red Army has treated them. It's time to surrender, give up their weapons. The war is over. He turned to the prisoners and told them they were free to go anywhere they wanted to go. They looked with hesitation at Korotkov and at the commander.

The commander just waved his hand in the direction of the forest. "Schnell … Go! Go now! Move!" said the commander. He turned to his own soldiers. "They are too scared and still can't believe their luck. They feel lost and can't make their minds up about what to do!" He turned back to the prisoners. "March on! Quickly! Schnell! Schnell!"

the commander said again to the prisoners. Finally, they followed his instructions. As they ran towards the forest, they kept turning back as if waiting for the Red Army open fire. Finally, they disappeared behind the trees.

"Do you think it's good that we let them go?" one of the soldiers asked. "We don't know what's on their minds."

"Letting them go is the only way now," one of the pilots replied. "The war is coming to an end." He was still trying to contact his squadron using the radio. Finally, he contacted his squadron. The commander-in-chief of the squadron was concerned because he had not been able to make contact, despite numerous attempts. The pilot explained about the night's fighting, the prisoners, and why the commander had decided to release them.

"Well done." Everybody heard the deep voice of the regiment's commander-in-chief. "Good decision, team. We have no further assignments. In the evening, expect the whole regiment to be at your airfield."

"Yes, sir!" the pilot replied firmly.

When the rest of the regiment arrived, there was lively discussion about the night's events and the decision to release the prisoners. Suddenly, one of the Red Army snipers burst into the bunker, shouting. "Comrade, fascists!"

As if blown by the wind, they all whooshed outside the bunker. Yes, without any doubt, from the far end of the airfield walking towards the planes was a crowd of men in ripped uniforms. To everyone's astonishment, they were walking unarmed directly towards the squadron.

"Look there's Stepan with them!" someone shouted.

It didn't seem real! But then everybody saw Stepan, a member of the squadron. He turned and said something to the surrendered soldiers, and all of them stopped. Stepan walked forward as everyone waited in anticipation. "Please let me report!" he said. "I've brought the prisoners. All their weapons are in working order and have been left in the valley."

"Explain exactly what happened!" the commander demanded.

"This is what happened, sir. Half an hour ago I decided to go to the far end of the airfield because from there I could spot the valley. I wanted to have a look around. As I walked to the edge of the airfield, about ten soldiers emerged from nowhere with machine guns in their hands. They grabbed me and pulled me to the bottom of the hill. I thought, *That's it. My life is going to end!*" he took a deep breath before he carried on speaking: "One of the soldiers, who was wearing glasses, came to me and said that they wanted to surrender. He said that one of their soldiers had come to them that the morning and told them everything that had happened when they were in the hands of the Red Army. I recognized him. It was one of the five prisoners we let go this morning. He was standing there and staring at me. So I said to them, 'Put your weapons down and get in line behind me.' I brought them here," he said straightening his body lifting his right hand to salute.

"Yep, now we have something to think about."

"What are we supposed to do with them?"

The next step was for a commander to contact the central regiment. This time he talked to the Red Army general, who said, "Do you know place called Kalay? Yes, not far from the motorway. About seven kilometres from where you are. Put them in the line and send them to Kalay by foot. Our boys will meet them there."

As the commander walked towards the group of surrendered soldiers, he ordered the one wearing glasses to step out from the crowd to the front. This soldier had been working at headquarters division and could speak fluent Russian, so the commander decided to hold on to him until the arrival of the general.

"Are all the prisoners in line?" asked the commander. Stepan stood at the front and, to be completely sure of keeping the prisoners in control, three other soldiers from their squadron were at the back of the line. The commander issued an order to relocate captives to the required destination.

The next day, Stepan reported that the line had been growing bigger and bigger, like a snowball, because many more had surrendered.

On arrival, each prisoner had been registered and handed over to be dealt with by the other squadron, Stepan reported.

While all this reporting was going on, Lina was sitting quietly in the corner wondering how her boyfriend was and when there would be the end to the horrid war. When would the suffering and pain come to an end? When would all of them be able to leave and return to their homes and their families? She wondered if it was just rumours or actual truth about the end of the war. She could not stop thinking about all of the questions in her mind. She was eagerly anticipating the end of the war. For her it meant a new beginning, a new life as she returned home and was reunited with her beloved family. She imagined how Peter was going to find her and how they could stay together forever.

The next day another regiment from Kalay joined the division where Lina was. Everybody was asking questions about the current events and the progress they had made. Many were satisfied about the enemy's surrender. It convinced many that the war was ending. In the morning the pilots were due to fly again. The order was to destroy the remaining fascist army.

It was the last days of Hitler's Germany. Red Army troops were preparing a battle for Berlin. One day a commander of the regiment Lina was assigned to assembled all the troops together to clarify the situations between different divisions. It was very important so the captains and everyone else were clear about the actions and interactions so that during the battle they would understand each other's tactics.

Apart from this, the commander had to hear from the captains of different divisions—infantry, gunners, and tank men. The whole of the battalion met the tank men and agreed to everything in the meeting. Then they all went to give orders to the front squad. They were the ones in the trenches who often became victims of unclear communication.

Suddenly, everyone heard music coming from where the last groups of the enemy were based.

"The enemy—initiators of death—are having fun!" the commander of the infantry said with anger. "They are drunk! The bastards are celebrating as if they're on some kind of holiday! The beasts are sensing the end of the war, and they think that their lives are going to flourish!" He was pointing his hand to his right. Lina and all the others looked in the direction he was pointing and saw the enemy raising a flag emblazoned with a black swastika.

The commander took out his rifle. It was clear that he wanted to shoot down the flag, but he changed his mind. At the same time, the commander was called in by General Ryazanov. "Yes, General!"

"That flag must be taken down. Understood?"

"Yes, General!"

"Go to the airfield and take an action! But bear in mind ..." The general spoke in halftone. "No one must hear you, apart from our troops. Take that flag down in one swift mission. This is extremely important. Clear? Go ahead!"

The commander rang straightaway to his regiment and ordered them to prepare the aircraft immediately. The Red Army pilots were in the air within minutes. Everyone knew it would not be an easy task, and the operation should be done swiftly in one move. But there was no time for further discussion. The general's order had been very strict. It was important that it be carried out with an immediate effect.

Lina watched. In an instant, where not long ago the flag had been flapping in the wind, the scenery changed into clouds of smoke. The assignment was complete, and everyone was satisfied.

Another difficult but completed task, Lina thought, praying for everyone to return in one piece.

"Well done!" The general shook hands with the members of the squadron.

"Very good job," added one of the tank men, and his mates agreed.

"With a team like this it's not difficult to carry out successful missions. We can do it, right?" said one of the pilots.

"Of course, brother," another pilot replied.

After few days, the division Lina was assigned to started to move out further to destroy the enemy. Tanks broke through the enemy defence with a helping hand from motorized infantry. The airplanes flew so low they almost touched the tops of the trees. Intense fighting continued throughout Germany as the Red Army stormed the cities of Gindenburg, Oppeln, and Oder.

Troops of the Red Army, including Lina's battalion, broke through the resistance of the enemy as they confidently and rapidly moved towards Berlin.

During the fighting in the region of RagovTeypits, a large group or German Army was surrounded. In the "pot" the Red troops discovered infantry and tank divisions.

In the early hours of the morning, gathering all their strength into one force, the fascists were able to break the surrounding ring and were going through the forest. They were trying to connect with their main forces. Also in the early hours of the morning, the task of the entire Red squadron was to find out in which direction Germans were moving. In the night, somehow the enemy had managed to vanish into the forest landscape. The Red Army's intelligence unit lost sight of them. This created some serious problems because an enemy force of considerable size was hiding somewhere ready to fight at any time. What was on their minds? No one knew, but the whole battalion was trying to think ahead and decide what to do next.

The pilots were searching through the forest at a height of five hundred metres trying to spot them. Lina was listening to the reports from the pilots. So far, they had not spotted anything.

The captain once again ordered, "Check the woodlands thoroughly once again!"

The pilots made another low pass, but they could see nothing. The ancient trees formed a solid roof over the forest floor. Everywhere was peace and quiet. So where had the enemy gone? They could not have just disappeared! Everybody was in despair.

Suddenly the voice of one of the pilots over the radio: "Captain, the enemy has opened a fire!"

Everyone's nerves were on edge. The captain spoke loudly: "Just fly very close to the ground."

"We can see a lot of machinery and tanks hidden in the woodland," the pilot reported so that everyone could hear.

"You must respond to their shooting!" the captain ordered.

"Yes, Captain!

"I'm going to send another group of pilots to relieve you," the captain informed him.

Lina listened to the captain's orders, and still her thoughts were the same as they had been when she arrived to carry out her duties. She was wondering where her Peter was and whether he was still alive and well. She knew a lot of troops had lost their lives in the last battles with the enemy. Lina hoped he was okay. She was dreaming that they would find each other again. Thinking this way helped her taking her mind off the horrid, long-lasting battles.

The front line was held at one hundred and sixty kilometres from Berlin. The regiment was located near a small town. Every day the pilots stormed the last of the major enemy's nests, helping ground troops to break through so they could destroy the remaining fascists. From a distance came the sound of artillery fire again and occasionally even the crack and rattle of rifles. This was the Soviet 17th Army pushing forward. The terrified SS were running.

Lina noticed that somebody had brought in a plan, and on it she spotted a map of Berlin. She thought that the war was coming to the end soon. Even the general had got all the pilots together to explain further plans of action. The Soviet Army was in full blast. Their forefront position meant that the war was near to the end.

"You must enter Berlin from the side of Luckenwalde slowly." The general spoke as if he was trying to choose his words carefully. "You must check the bridge on the western side of the city. Then check Potsdam. Look at the situation there. Then come home. Do you understand your task?" he asked the pilots.

"Yes, General. Everything is clear!" answered one of the pilots.

This flight was not easy. In fact, it was quite risky. One of the pilots had to fly straight into the enemy's centre point. It all had to be done in a day too. Every pilot knew that Berlin was guarded by anti-aircraft artillery. Even at night bombers had been lost.

Lina put the cards down, turned, and addressed the general. "I would like to ask you something."

"Please do, Lina."

"This flight is very dangerous, and I'm worried. I don't want my colleagues risk their lives!" she said firmly.

"Why do you think this?" The captain came closer to Lina and put his hand on her shoulder. He spoke to her in a fatherly way. 'Everything will be fine, child. We are waiting for a report from our spies. Don't worry, Lina."

Lina got up from the chair and went outside. She was concerned for her colleagues. She did not want them to be in danger now that the war had nearly come to an end. She was trying to be protective, but deep inside she knew no one was guaranteed safety.

In no time the pilots were inside their planes heading for the sky. Lina went inside to listen to the radio transmission.

"I can see the bridge over the river," she heard the pilot's voice. The pilot was waiting for the order to attack. "I'm going to attack, now!" The pilot said impatiently.

But the general banned the attack. "Cancel the attack! Don't attack!" He firmly commanded.

Lina heard on the radio explosions and shots. Suddenly, radio contact with the pilot was broken. The general couldn't contact the pilots.

"I can't contact the pilots!" the general said. "Let us believe that the boys know what they are doing and will do the job well," he said as he walked out of the bunker.

Waiting in anticipation, Lina and the others heard the sound of aircraft. "One, two, three, four ... five ..." Someone was counting loudly.

"Where are all the others?" asked Lina. That was the first question that everybody wanted to ask.

"Mission completed, General!" reported one of the pilots.

"Why didn't you listen to my command, Gavrilov?" the General asked sternly.

"Sir, at the time I had a clear view and used my military experience to hit the target directly!" he replied with a total confidence.

"Okay. Where are the rest?" the General asked.

"We lost them," Gavrilov replied. "They were shot down by the enemy, General!"

After some silence, the general looked up, and with his stern gaze said: "Good job! Well done! Go get some rest!"

"Yes, General!" The pilots saluted, lifting their right hands in unison.

When the pilots left the bunker, Lina and some of the others from the regiment waited outside in anticipation, wanting to hear what had happened. Lina, as always, spoke first. "Tell us, guys! What went wrong?"

One of the pilots stopped. He sat down, too tired from completing a very complex mission. He began telling everyone about the air fight. "I was in the sky," he began, "listening to the general's commands. I could see the roofs of the houses, the fire towers, and the churches. The fascists couldn't shoot me down because I was flying too low, so they shot at the back of my plane just for the sake of shooting. Underneath me flashed the images of streets full of people, machinery, buildings, and ruins. Everything seemed quiet, but it was only because I was high up in the sky. The fascists were full of confusion and fear. They knew the collapse of what they had started was imminent. Sooner or later they would have to answer for the crimes they'd committed. They were twisted from the fear, clenching their teeth with helpless anger.

"Anyway, I had completed my first task, and then I saw Potsdam. The enemy were running away. They were following me, and then I saw a huge fire all over Potsdam. So I decided to come into it from another side. As I turned around, they tried to shoot me down. I had lost contact with the general too. I wondered if it would be right for me to leave without carrying out the general's order." He paused. "No, I had never done anything like that in my career. So I kept thinking. After several minutes, I made my decision. I increased my altitude. I knew it was dangerous, but there was no other way to do it. I expected that, at any minute, the enemy fighters would appear. I circled the area, trying to observe the positions of the enemy artillery. I was going to go for it again before returning to the airfield. Again, they launched an attack with machine guns. I repeated the course to compete the task. When I looked around again, I couldn't see any fighters. I wondered if they'd run out of fuel. That has happened to them quite often lately, according to the pilots' reports. At last I could hear the general's voice. He let me know that the other pilots had managed to attack the airfield. All the enemy's machines were in flames. I knew our guys should be close somewhere, and I was looking for them. But no luck. then I recognized one of our aircraft lying in the enemy's airfield—all in flames!"

When Gavrilov finished, there was silence. Nobody could say anything.

After one day of rest, the commander of the regiment came up with a new task. They were going to Berlin.

The pilots stood and listened to their orders. The task was to get through to the centre of Berlin and find a barge full of tanks. Intelligence services had already done their job and reported the position of the barge. They were ordered to sink the barge so it couldn't be used.

"The tanks are Hitler's own reserve," explained the commander. "Are you clear about this order?" The pilots confirmed by nodding.

"The entire city is full of the enemy's tanks and vehicles, so why allow them to have more? This should not be allowed."

"Twenty-four machines are ready to fly right now," said Stepanov. "They are from two squadrons. But please let only our squadron to do the task!"

"Why?" the commander asked curiously.

"With fewer planes, we will make less noise and, most importantly, we will sustain fewer losses. It will be easier for them to target twenty-four planes crowded together. Each shot will get at least one plane. With only ten planes in the air, the chances that they will hit a target on the ground are much better. Also, it would be too crowded having twenty-four airplanes circling in the air over a barge. The target is too small for twenty-four."

"I see," said the general. "Do whatever you think is right."

After several minutes, twelve aircraft rapidly took off, one by one, from the runway. In the air, they lined up in a triangle formation heading towards Berlin. Now the route was no longer dangerous because the Germans had gathered on their last piece of their land in Berlin. Berlin was not too far from the airfield from which the planes had departed.

The shooting started when the airplanes reached the city. There was continuous fire, but the pilots knew what to do.

"We see the barges! There are two, commander!" Gavrilov reported on the radio. "The tanks are uncovered. Looks like they are being prepared to be taken off the barge. We are going to attack!"

Everybody could hear over the radio the sounds of shooting, explosions, and blasts!

"The tanks are sinking, commander," the pilot reported.

"Well done. Carry on. When the mission is complete, return to the airfield. We will be waiting," the commander ordered.

"Yes, sir," the pilot replied.

At the airfield, Lina and all the squadron members waited impatiently for the pilots to return. The mission had been completed successfully, and most importantly there had been no losses. All squadron members ran to the airplanes as they landed. They pulled the pilots out of the planes and hugged them, slapping them on their backs.

"Congratulations and thank you, from all of us!" said the captain. "On behalf of the general and all the other commanders, we would like to express our gratitude! Nobody will ever forget your awesome bravery. You must be proud, lads!"

Lina was relieved that this time everybody had returned in one piece. No one had been injured. She knew the next day would be the beginning of the storming of Berlin. She was scared, but at the same time she had been waiting for this because the war would be over and she would finally be able to return home. This was what she had wanted and what she had been dreaming of and thinking about all the time. This was the reason she had been determined to survive against all odds. And now that she had a man in her life who was so dear to her, she was determined to see him again. Nothing would stand in her way and break her now.

Half of the city of Berlin was covered in ruins. The fascist army had been broken, but they continued to fight virtually under the guns of their commanders. And Hitler's leaders tried to put at stake the city and its habitants. But it didn't matter to the fascists now. They armed women and the elderly with machine guns and gave guns to young boys who didn't even know how to hold them properly. These people were under certain death sentences. Many died during that appalling week, of cold and exposure, of exhaustion and malnutrition, of wounds old and new inflicted by the fascists' ill-minded beliefs.

The pilots left one morning to complete orders assigned to them by the general. First thing in the morning, Lina was prepared to hear the reports from the pilots about the situation in the air. The entire squadron was in the air, storming the remaining fascist beasts. Lina heard the regiment commander's voice on the radio. He spoke clearly without any hesitation whatsoever. "So, who should we be afraid of? Over. Anybody?"

"Gavrilov! Gavrilov! In the area—German tanks. Lots of them. Help! Follow me!" Gavrilov ordered the entire squadron to turn around and help!

The commander's voice was heard on the radio once again. "I see tanks, but not many," he reported.

"Pavlov, report! Pavlov, report! We are surrounded by Germans! Help us, pilots! Strike my tank if you have to!"

Gavrilov looked harder and saw several German tanks. "What are you saying, Pavlov? Repeat! Pavlov?"

"I'm saying strike now!"

"Attack now!" Gavrilov commanded the whole squadron!

The commander's voice on the radio confirmed the order. The pilots targeted with precision and accuracy so they would not to damage the Red Army squadron. Seven tanks had been bombed and stopped, and two of them were reported to be on fire.

Another attack with an accurate precision struck the enemy machines. German tanks that were only slightly damaged tried to escape by getting lost in the trees! This was the moment when the pilots had the same idea—to forget all thoughts of humanity and destroy the sick bastards.

Another successful attack! The war was ending. Everybody knew it, and in just few days everyone would hear about it.

The tank regiment arrived in the evening with the general. Once again, there were happy greetings, handshakes, and laughter. The general raised several toasts to the pilots' well-done job. He commented about the importance of friendship between different regiments and how invaluable their work and tactics in targeting the enemy's military assets had been.

"I would like to express a big thank you for saving my life in the last fight! To Pavlov!" The general rose from his chair with a drink in his hand. "Soon—very soon—brothers, we will be toasting our victory! But for now, we will raise our glasses and toast the bravery and good work of our soldiers!" He drank his drink in one go! "It was just a superior attack! We were surrounded, and then you guys came and hit our enemy right on target! Well done!"

After dinner, everybody was in such good spirits. The usual tension that had been hanging in the air was gone, and Lina enjoyed seeing the happy faces. Everybody was talking about the end of the war. Lina was happy, thinking about her return to the Ukraine! She was dreaming about the moment when she could see her family and see her lovely Peter. What she would say when she saw her mum and dad and Ivan? There was so much to tell! So much had happened since she had been forced to leave her home.

Now she was thinking about Peter. The last time she had received a letter from him was two months ago. He had told her that he wanted to see and hear her and that he loved her and could not wait for the time when they would be together again. Lina dreamed about him. In her dream, Peter took her hand and asked her to marry him. She woke up!

After helping all day with the injured and assisting the evacuees, Lina was sitting near the fire listening to the singing voices. She was tired from hard work. Lina had become friends with everyone. She was like sister to all the members of the regiment. She was valued for her kindness, modesty, and simplicity. When the fighting started, everyone in the regiment had been astonished at her self-discipline, independence, bravery, and organization abilities during extreme situations. Her good character shone through as she showed love to the injured while she tenderly treated and cared for them. She was respected by all.

Now she sat by the fire. It was the last night they would spend together before they all went separate ways. She was animated and sociable, but there was something in her that seemed to be odd, and this was reflected in the way she behaved. There was something else playing on her mind and in her eyes. She seemed to be hiding something that could not be explained in words. And when everybody started to say their goodbyes, Lina suddenly burst into tears. This was totally unlike her usual behaviour. Everyone was surprised. Finally, she calmed down and made everyone promise that they would not tell anybody else what she was about to tell them.

She began her story: "A few days ago, I was helping the injured during the battle. The investigation was carried out during the first half of the night, in wind, heavy rain, and poor visibility. The fight was unsuccessful. We had evacuated the wounded to the rear. I came into the bunker. There was a heavy feeling in my heart. I asked one soldier, then another soldier, but nobody knew where the commander was.

"Finally, his whereabouts were clear. At a critical moment of the battle, to rectify the situation, the commander had collected a few fighters and determined to set example for the soldiers in the attack. Some saw him fall about hundred metres from the enemy trenches. Soldiers were lying on the ground trying to avoid the deadly fire, and they started to crawl back. The lieutenant stayed there.

"I automatically realized that the wounded commander needed my help. I decided to look for him and rescue him. But how? How could I lift him? I got on my knees and begged other soldiers to crawl with me and find the commander. But nobody agreed to help me with my rescue operation.

"One of the surviving officers in charge, for fear of liability, did not want to enforce the order to rescue a lost commander, as he knew he would be sending someone to die. He also hesitated because, in such weather conditions, it would be almost impossible to find him in the field of muddy sludge, or to bring him back if they did find him.

"I decided to go alone. Time was running out. I started my rescue operation. I took a number of packets of bandages and put them inside my uniform and crawled out of the trench." After a short pause and wiping her tears, Lina carried on: "It was after midnight and very dark. Cold rain poured down as if from a bucket. The tension of the fight had lessened, but the rockets continued bombarding the land with loud sounds. Glowing bullets continued to whistle overhead. Lightning rockets were flying over the German trenches. I would keep still for a few moments, look around, and then crawl again. I must admit," she continued, "I was dreadfully scared, but I continued to crawl. My boots were full of muddy water. My hands were covered in the mud. The tears were blurring my vision, but I kept crawling …" She paused. She was only eighteen and had seen more horrors than most people do in a lifetime.

"What made you crawl on your own? What made you do this?" asked one soldier, shaking his head.

"I'm trained to carry out rescue operations. Saving soldiers is my duty," she responded, and then she continued her story. "I kept on crawling metre by metre moving closer to the enemy's trenches. This war is very brutal, constantly requiring bravery, sacrifices, and losses from all of us. Perhaps I've turned against logic because I could not do it any other way. It was my duty to help the vulnerable injured soldiers whose lives were in my hands."

As Lina spoke, the faces of the soldiers became serious. She could feel with her female instincts that everybody who was listening felt everything she had gone through during the rescue operation. Some other rescuers from squadron knew exactly what it was like from their own experiences.

"I was so scared, but I kept on crawling," Lina continued. "A rocket would fly, and I would stop in silence and then carry on. Automatic machinegun fire flew over my head, and rockets fell nearby. I kept on crawling. I could feel the bodies of dead. Some of the soldiers told me that the commander had fallen near the enemy trenches, so I thought that maybe he was crawling out and might be somewhere nearby. Still, I carried on crawling, guided by the flashes of light coming from the rockets. That light was the only relief from the dark. Suddenly I could see something moving in front of me. I was careful because it might be an enemy soldier who had left the trenches to look for our wounded soldiers.

"So I thought that my life was going to end right there. I wanted to shoot, but I changed my mind. I couldn't even hold a gun in my hand because it was numb from the cold mud. With my teeth clenched onto the ring attached the pin on a grenade, I lay very still and waited for the right time to pull. I saw someone crawling right towards me. I counted the metres in my head. I made a decision: One more metre, and I would pull the pin. In my mind, everything just stopped. I began to think that I wasn't going to give up. I tried to calm myself down, but I still held hard onto the ring of the grenade.

"Suddenly, the enemy soldier decided to turn in another direction when he was about a metre away from me. Maybe he hadn't noticed me, or he had taken me for a dead soldier. I had not moved, but the fear in my heart turned to relief because I was still alive. My heart was beating so hard, I worried about the enemy spotting me. As soon as my breathing returned to normal, I carried on crawling.

"I was scared, but I had to complete the rescue operation. I crawled up onto a small hill, and a rocket lit up the sky. I found the lieutenant. I recognized him because of his wide belt, which shone slightly in the light of the rockets. I crawled to him and turned him over. When I put my ear to his chest, I could hear a weak heartbeat. His whole body was sunk in the mud, making it almost impossible for me to move him. Somehow, I managed to get him out and started dragging him across the mud. When I had crawled about hundred metres, I felt drained. I took two packets of bandages from inside my jacket and quickly bandaged his wounds. I unwound the rest of the bandages and looped them under his arms. Then as if I was a horse, I started pulling him behind me. It was the easiest way to move him. I would pull and rest and then carry on. I don't know how long it took me. With whatever strength I had left—and it's astonishing how much strength you still have when your life depends on it—I tried to get back with the lieutenant no matter what. I was doing it automatically, crawling from the enemy's trenches. I carried on crawling until I fell in to our trench. By then, all my strength had completely left me.

"After what happened, I think I must have passed out. I heard voices from distance. They grew louder, and I realized that someone was shaking me by the shoulder. I cannot remember how I was brought into the bunker and wrapped in the warm blanket. I slept all day. When

I gained consciousness in the evening, I tried hard to remember all the events, but I could not."

When Lina finished telling her story, everybody knew why she was respected by so many. Without a doubt, Lina's bravery caused her to be admired by her squadron. In their eyes, she was their little hero.

In two days, Berlin was taken. Reichstag—the parliament—was disbanded! On the seventh of May, the pilots of the 144th squadron made one last flight to Berlin. On the eighth of May, representatives of Germany signed an act of an unconditional surrender. This news came to the regiment two days later. When it did, Lina woke up from the loud noises. She quickly got dressed and ran outside. Some people from her squadron was on the airfield. Pilots, mechanics, armourers, soldiers of defence, and airfield servicemen were running all over the filed throwing their hats in the air, embracing each other, and shouting. Lina could hear one word clearly. The word was *victory*!

One soldier ran towards Lina, grabbed her into his arms, and squeezed tightly. "It's our victory, Lina!" he shouted! "The war has finished! Victory! Victory!"

Lina finally understood. "Victory!" she cried. The sudden burst of energy made her deliriously happy. She screamed and shouted. She was so relieved and so glad that the dreadful war was at an end, and she was looking forward to the beginning of a new life.

As Lina celebrated, her feet tapped on the ground leaving shoe prints in the dust. Her hair was shining, and her face was lit up with a smile. She hadn't felt so happy for a long time. She felt so happy, in fact, that she raised her arms and moved her feet in a rousing dance—a *chechetka*! This was her victory dance.

Tables had been set up. The radio was transmitting the latest news. Across Germany a de-Nazification programme was rapidly instituted after the war. Many Nazis had already abandoned their uniforms, slipped off their swastika armbands, and walked quietly away. Adding to the news, the men at the top of the Nazi regime, who had engineered the whole thing, met their ends in ways that could never justify the millions of deaths they had caused. Adolf Hitler, consistently voted even today as the most evil-minded man in history, took poison or possibly shot himself in the private apartments of his bunker under Berlin as the Red Army battled its way through the streets of the capital. That happened on 30 April 1945.

The streets were filled with cries, shouts, and yelps of joy. The war had finally come to an end. There were happy celebrations on the streets of Germany on the day that each and every one had been waiting for and dreaming about. Finally, they had come to the end of the long-lasting war. Now all of the soldiers and captive slaves could finally return to their families and try to rebuild their lives after so many losses brought about by the brutal realities of the Second World War. These survivors had lived through a chaotic, bewildering, confusing situation designed by the Nazis.

How could the youngsters grasp the context of war, the Nazi propaganda, the brutality, and Hitler's twisted mind? Not until after the war would the world find out that nearly half a million people had been gassed during this period. Today the statistics show that, in two days and nights in July, 7,000 people from the family camp were murdered and thrown to a mass grave. Each one of these people was a hero—the men, women, and children are honoured by many who live today and remember the harsh realities of the war. It is hoped that people learned from it so they will not to repeat the costly mistakes. It is hoped that people will work to eradicate evil-minded leaders who brainwash others into false beliefs that promote misery, devastation, and loss of life. These beliefs steal the most precious of God's gifts.

PART 7

Return Home

CHAPTER 20

It was a bright sunny day, and people were returning to their various homelands. It was a time for rebuilding lives after the destructive, long-lasting war. For many it was a time of new beginnings that would set them on different roads chosen by fate.

Lina was on the train heading to Zhytomyr in the Ukraine. The wheels of the train were making squeaky noises, and the rays from the sun were playfully appearing from time to time through the boards that were the roof.

She was imagining what it would be like when she saw her dear dad, mum, and little brother. So many years had passed. She didn't know how they all were. She had not received a single letter. She had received no news from the moment of that dreadful night when she was taken.

Many nights over the past years she had thought about the day when she would return, and this was the day. It was the moment she had not only waited for but also dreamed of. This was the moment that had kept her going despite the hard work day and night. The journey back home was long; it seemed endless.

The train has finally arrived at Zhytomyr. Lina's heart was pounding. She could not believe that, in just a few moments, she would step onto the soil of her homeland. Slowly stepping from the train, she looked around and breathed in a gulp of fresh air. The surroundings and the landscape warmed her heart as she stood on the lonely platform. There was no one else there; she thought the station was empty. But then she noticed someone else on the same platform. It was a woman, and she appeared to be looking for someone. The women began to slowly walk towards her. "Good evening," she said to Lina. "I couldn't help but notice that you must have come a long way. I can see you have heavy bags."

"Yes." Lina nodded.

"Don't worry," said the woman, sensing Lina's caution. "What village are you from?"

"Babrichi," she replied quietly.

"Don't go there through the forest, child," the woman warned. "It's very dangerous. The local people have heard all sorts of stories that would send chills down your spine. The other day someone got robbed. Some people were seen getting off the train. They went through the forest, and no one has seen them since. Bad people are in the forest," she said with a concerned look on her face. "It's getting dark, child. You can stay in my house for the night. I've got a daughter just like you. That is why I am here now, and she is due to be here any minute. She works here at the station, and I'm waiting for her to finish her duties for the day. It seems that she is running late today." The woman smiled.

Lina desperately wanted to see her family. It was so frustrating knowing that she was only a few kilometres away from them. But she decided to stay the night with the two women. After all, it would be safer and would give her a chance to relax after a long journey.

The women lived not far—just a short walk from the station. "It's only two of us, me and my daughter," said the woman. "Come inside, child. Give me your coat. You must be hungry travelling for so long."

"Yes," Lina replied hesitatingly.

The woman's daughter set the table and poured soup into wooden bowls.

"Smells nice!" Lina said. It reminded her of her mother's home-made soup. It was green borsch cooked from nettles. Lina sipped the watery mixture, and it warmed up her stomach. She was feeling safe and welcome in the nice woman's house.

The woman sent a telegram to the village of Babrichi. Lina sensed a compassionate, caring, motherly warmth from this woman. It felt good to be back. The woman hadn't wanted Lina to walk through that forest on her own.

The news travelled fast. Ivan was the first one to hear. In no time, he was in his cart. He drove through that forest like a bullet. In no time he was in the front yard of the woman's house. Lina opened the front door. Ivan grabbed her tightly, and they burst into tears—the happy tears of togetherness. It was an emotional reunion with her brother, whom she hadn't seen for years.

She asked as if to reconfirm, "Is it you? Is it you I'm holding?"

"Yes, sis, it's me!" he replied, laughing. He kept holding her hands, too scared to let her go. "You're back, sis!" They were laughing together, so happy.

"I'm back!" she proclaimed with joy, happy tears, and the warmest of hugs. She was with her brother at last. It seemed so surreal that, after so many years, they were reunited. They had survived the most horrid war in the human history.

Hugging each other, the siblings experienced such happy emotions. They were so dear to each other. They both cried, not wanting to let go of each other. It was one of the most triumphant moments in Lina's life. She had only been able to dream about this several weeks ago; now it was reality. It was happening. She was ecstatic—overwhelmingly happy.

Lina couldn't thank the women enough for their hospitality. She gave them a few of her best dresses before she left. As she and Ivan headed in the direction of Babrichi, the loud snorting of the horse and the jingle of its bridle and halter sounded so familiar and yet so unusual at the same time to Lina. The horse obeyed Ivan's commands and took them with incredible speed to Babrichi. It was almost as if the horse sensed the siblings' urgency.

"Brother, you don't know how glad I am to see you!" Lina was so content and very happy.

"Why didn't you come home sooner?" Ivan asked.

"I had to work for the Red Army. It was compulsory, and it was the only way not to be accused of perpetrating the country. If it had been up to me, I would have returned sooner, but it was out of my control. I had to work hard till the fascists were completely defeated. My work was intense and difficult, but much needed in the army because of the enemy's tactics. They thought they could take over, but our army showed them where crayfish were hiding in winter.

They both laughed. They *could* laugh now. It was their right.

"I thought about you all the time," she told him. "I dreamed about the day of my return. Brother, it was horrific. I've seen so much—men, young and old lost their lives, as did women, boys, and girls." Her eyes widened as she spoke. The lines on her forehead indicated every tension she felt. The horrors she'd been through were evident in her facial expressions.

"Far from home, young, and inexperienced, I learnt how to defend myself—how to hold and shoot a rifle, how to clench my teeth and fight for my survival. When I and the others heard about the victory, I couldn't believe the news. It had been such a long wait. It was the moment everyone had been waiting for! Everyone was shouting triumphantly: Victory!"

"I know, sister. When we heard it on the radio—Oh! All villagers got together to celebrate the happy news!" he said with a huge excitement.

Soon they arrived in Babrichi. Lina looked around. The war had caused some ruin and destruction, but the beauty of the charming countryside—the fields, clear skies, singing birds—were signs of peace. She couldn't believe the extent of the destruction; some houses had been knocked down by the bombs. The little cottages that still stood seemed a little smaller than she remembered. In fact, everything looked smaller. Not many people were outside. In the days of her youth she would see children playing and women and men getting on with their jobs. Now the village looked silent as if the angel of death, wearing a swastika, had visited Babrichi and swept everything and everyone with its "magic" destructive wand.

Lina knew that the enemy had been defeated, but she was shocked that the invasive nature of the war had left such scarring and destruction on the familiar landmarks. Taking a deep breath, she let all the tightness out. She breathed it all out with a sigh of liberation. "I'm so glad to be back!" She proclaimed with positivity.

The cart stopped outside their house. The fence was no longer there, but the little cottage had survived the war. Lina jumped off the cart and walked towards the house, bending under the windows so no one inside could see. Carefully she crept to the front door of the family home. Opening the front door wide, she was surprised to see her cousin. "Lina!" Katya greeted her with open arms, hugging and kissing her. "If you had arrived just a little bit sooner, you would have been able to see your mum," her cousin said.

Lina's body froze. "Has my mum died?" she asked quietly.

"No, of course not. Silly me!" Katya was laughing. "I must have said it wrong. Aunty Yarina went to the river to wash the clothes. She should be here very soon!" She laughed again. "I nearly gave you a heart attack, didn't I? I'm so silly." She giggled. "Come on. Let's unpack your clothes. I've prepared some dinner. You must be starving." Katya, as always, was so chatty. Nothing could stop her talking. "Aunty should be here any minute now. Oh, she was waiting for you. She just couldn't sit still because of the excitement of your return, dear!"

"Huh! Thank God Mum is okay. Where is Dad?"

"You should have seen your mum's face when she heard you were here!" Lina's cousin kept on talking.

The minute her cousin finished her sentence, the front door opened wide. Lina's mum walked into the house. Lina run into her mother's arms, almost lifting the old, fragile lady off the floor. She was delighted to see her mum after so many years. No words could describe the blessings of one who had survived such a horrid wartime journey.

"My dear, lovely mum!" Lina cried with happy tears. Yarina was holding her daughter's shoulders as if she would never let her daughter go again. "I missed you so much, Mum." Her mother's arms wrapped around her, hugging her tightly. "I missed you so much, but I never lost faith that I would see you again.

"Come with me," Yarina said, holding onto her daughter's hand. "You have become even more beautiful." She looked at Lina with so much love and affection.

"Mum, I'm so glad to see you!" Lina kept repeating. She gently squeezing her mother's hands. "You look the same," she said. "Not a single day went by when I didn't think about all of you. Where is Dad? He's working! Isn't he?"

Everyone looked at Lina. She didn't like this silence. "Where is he?" She repeated her question again.

Her mum spoke first. "Daughter, your father didn't comeback from the war."

"What? How? I was looking forward to seeing *all* of you! No!" She sobbed over the loss of her dear parent who had always meant the world to her.

Katya comforted her cousin with a hug. Ivan put his hands around both girls. Yarina also comforted her children with motherly love and warmth. Their little family was grieving as they held onto each other with loving loyalty.

Everybody in the family was grief stricken. Maxim had always been a good husband and father. It was too painful to lose such a great man who had always made sure his family came first. Now Lina knew she was never going to see him again. It was too painful for her to take in. The love that she had for her father was immeasurable. Lina would have to learn to cope with the tragic loss of her beloved dad. Only time would heal the wounds of irreplaceable loss.

Yarina held her daughter's shoulders and walked her to the bench outside. They both sat down, and she began telling Lina what had happened. "After you were taken, there was an unbearable tension at home," she began. "Everyone was down. Your father felt so helpless when he realized he couldn't do anything about it. After the SS took you that night, your father wasn't himself. He became very agitated, angry, and restless. Every day we cried; we were so worried about you. It was as if a dark cloud had covered the sun forever. One morning your dad got up and told me that he was thinking about joining the army. I could understand that he wanted to free you, and I could not say a word. He volunteered to serve, but no one would take him because he was too old. They all thought he must be mad. Despite all their comments, he insisted. He said it was very important for him to join because his daughter had been taken. I knew your father. He would have done anything for you, and once he made up his mind, no one could convince him otherwise."

Lina hugged her mother, and Yarina continued her story. "Your father told the recruitment officers, 'How can you not understand me? I cannot just sit and wait while my daughter is out there in the hands of those cold-blooded killers risking her life daily.' He asked them if they had children. He told them that, if they had children, they would understand what he was talking about. 'Sod the age requirements,' he told them. 'I want to join the army!' He insisted, and no one could convince him otherwise, not even me. Your father was very persistent. In the end they took him into the army," she said with a saddened, tired face.

"He wrote letters to us," she continued. In his last letter he wrote: 'Yarina, I'm going to the last battle. If I survive, I should be home very soon.' He didn't return home. We received documents with blood stains and a bullet hole," Yarina said as tears coursed down her cheeks.

The war had turned Lina's mum into a widow. There were dark circles under her eyes, her body was fragile, and her hands were wrinkled and twisted. The hard work over the years in the government farm fields had taken it's toll, along with the starvation of the war. Yet she still had plenty of life coursing through her old body and love in her heart for the children who needed her. She wiped her face with her apron and continued telling Lina about her dad. "When the men from our village returned, they told us that the front had been moving with the victorious power of the brave soldiers. One day they had been sitting in the trenches when they were attacked. That was when your dad climbed out from the trench. He began was shouting, 'My daughter! My child is here somewhere!' He fought and ran to the front. The men said everyone had to run, and no one could stop in the attack." Yarina and Lina both were silent. After a moment, Yarina sighed loudly. Taking Lina's hand in hers, she said, "That was how I learned that your father was gone and he was not coming back."

Lina's face was wet with tears, and her heart was filled with pain. She couldn't say a word. The shock that she would never be able to see her father was another terrifying ordeal she must endure. She would never be able to get used to it. Only time would fade her pain and heal the wound, but the scar would remain. In her mind, she remembered the last time she had seen her father. The train had begun to move, and she saw him standing amongst other people—parents and relatives—watching whilst the captives were running and being shoved violently into the train cars. He was very dear to her, and her love towards him would always remain as a sparkle in her heart. He would never fade away from Lina's memory. Her father had been the best. He had been her rock, her protector, and her hero.

The painful news would stay with her, but her love for her father would never vanish. He was the only man who could take care of his family properly, and now every bit of work had landed on Ivan's shoulders. He had been obliged to take over the family.

Having dinner together was good, and for the first time after so many years, Lina felt the warm feeling of home. She truly enjoyed her mother's home-made food. She listened to stories about how life had been in the village during the wars. She learned how fascists had robbed the villagers, taking their possessions, destroying their homes, taking away their farm stock as well as their children. They took machinery, equipment, and horses from the farms. Members of the SS commando unit burned everything, including people, as they tried to hide their crimes. When people began to receive news about the friends and family members who had been killed in the war, everybody was supportive and united. They helped each other as well as they could with the very little they had. Every crop was gathered and given to the army, her mum told her. Children worked in the fields. Times were harsh. "Our village had a field of rye grain," Yarina said. "About five kilometres. It could feed the villagers, so all of us worked on it, from young to old. One morning Nikolay Nazarov—you know him? He was the only man in the village. They didn't take him into the army because of his ill health. He warned us that the Germans had landed near the river. He told us that all the villagers should be on guard. We should hide any food we had and protect the field of rye. He told us to go where the enemy was and destroy them. We asked how we could go there empty handed without any weapons. We took the tools we used for working in the fields and followed our leader. Five kilometres we walked, tired but not scared. We were prepared to fight to defend ourselves. When we approached the gully that separated us from the field, we saw something dark among the neat piles of rye. Our commander immediately declared, 'That's them hiding! I'm sure!' he shouted.

"There were forty of us, all women. Cautiously we managed to cross the small river. We walked up the hill, and without any hesitation we walked towards the enemy. Suddenly we noticed in the field an enormous cluster of rooks eating the grain. They were filling up before flying south. We all just burst out laughing, but our leader was red faced from embarrassment and told us that someone had started a false rumour. He promised to bring us some food and left. Very shortly afterwards, he brought us milk and bread. It was a hot day, and we sat down next to the heaps of rye to eat our lunch. Suddenly, our leader, looking into the distance, again shouted: "Here they come!"

"Looking closely, we saw three men walking towards us. They wore uniforms with badges and caps, and one had what looked like a rifle hanging behind his back. Our bread stuck in our throats from the fear. We all froze, and our leader ordered, 'Get ready!'

"We were fully prepared to fight our enemy once again even though we couldn't really see the people who were walking towards us. Suddenly one woman, Nastya Uhimova—you know her, child—shouted, 'It's my son! Oh, dear God, I didn't recognize you!' She started to run, opening her arms. Well you know her—she limped more than she ran. He grabbed her tightly. The mama was hysterical with happiness. We all cried. Even our leader's eyes watered. It was a group of young men returning home. The boys had been allowed to go home. All of them were from the Zhitomirskiy region and had been transferred to fight. It turned out that Nastya's son had a guitar, not a rifle as we had all thought at first. Our leader took it from him and tried to play a happy tune to celebrate the moment of triumphant return. This experience made us laugh. We hadn't laughed so much for such a long time."

Lina listening to her mum as she told her what had been happing in Babrichi. "During the war our village lived under a motto," said Yarina. "All for the army. Everything for victory!"

She continued, "When, in 1942, they began to collect funds from the Szitomirskaya oblast (administrative division) for the tank divisions, everybody contributed. That was a strong defence for the Soviet front. The patriotic gesture of our farmers was glorious confirmation of the inviolable unity of our motherland. We were giving away clothes, food, and tobacco to help the injured soldiers. Ivan and I became blood donors on a regular basis. They drew the blood from my lumpy old veins!" She stretched her hands, showing them to Lina.

"The summer of 1942 was a very difficult and dangerous time, especially when there were enemy airplanes bombing railway in Zhytomyr day and night. It was horrific, and we were scared to leave our homes. We would hide in our cellars because there we felt safer than in our houses. I prayed every day that they would leave the villages. We lived each and every day in desperate fear. The year 1944 passed and the new year, 1945, began. We were listening to the radio as news was transmitted. We were hoping for the end of the war. Finally, in May 1945, the news reached us. That was an unforgettable day. The soldiers began to return home. Young men and women were returning from SS captivity. Only God knows how many days and nights I prayed for you to return." Yarina wiped her eyes with her apron as she told her daughter how life had been for those left in the village. Lina comforted her mum, hugging her fragile, thin body and holding her wrinkly but warm hands. Katya and Ivan were listening, so captivated by the story that they didn't even a blink. Four survivors around the table had lived through the tragedy and fought their fights, yet they all felt blessed with the gift of life in their small cosy little cottage.

CHAPTER 21

Life in the village was hard as Lina tried to forget the terrible years of slavery, hunger, and devastation. It all seemed so recent in her mind. Very often she would have nightmares with terrifying flashes of transportation, camps, and the commanding voices of the SS men. The scars of war were still too fresh in her mind. The most important thing that mattered to her during that time was that she was with her family.

Lina was extremely busy. She worked on the farm during the day and helped her mum in the evening. Indeed, the farm occupied most of their time. The land needed to be worked, and they had had to do it all with their bare hands. They had to collect the harvest just before the winter season. Spring, summer, and autumn were the busiest times with so much work to be done and only three of them to do it.

After Maxim left, Ivan had to do all the man's work. He was the only man in the family, so all the jobs that Maxim used to do became a heavy burden for his son. At the time, he was a young lad with the responsibilities of a father. Maxim used to do all the heavy jobs, which made life easier for his wife and daughter. Ivan's duties included collecting wood from the forest to keep the house warm. The young boy had inherited many skills from his father and could do almost everything on the farm to help his mum and sister—until one day a telegram arrived from the main regional government office stating that all the men in the village had to go to work for the government. It was a compulsory request made by the local council. The men's labour was needed everywhere in the post-war Ukraine to build new roads and railways and other new infrastructure to replace what had been totally destroyed. When Ivan told the news to his mother, Yarina just fell to her knees. She was a poor and honest woman, tired from the war and its tragedies. It was a heart-rending scene. Lina watched her mother and Ivan, both in tears, standing in the middle of the room. Her mum was screaming, clenched her son's shoulders with her fingers. She did not want to let him go. "Why now?" she cried. "Why do they want to take you now? Why do they want to take you away from me?" Yarina said asked the same questions repeatedly. "My daughter has just returned, and now my boy—the only son I have—must go!" She was sobbing, too afraid to let him go. "Why can't my children be here with me?" After a while, it was as if something had become stuck in her throat. She was so upset, she couldn't even talk. Lina tried to say something helpful, but the words would not come. She wanted to comfort her mum, but her body completely froze.

"Mum, Lina, I am going to return!" Ivan exclaimed. "I will earn some money so we can grow more crops and get our farm going. You have to promise me that you won't cry." He gave them both a stern look.

"Yes, son." Yarina wiped her face with the bottom of her apron.

Ivan finally felt that he could leave.

Not obeying the government's orders was punishable in the communist post-war society. The country needed to be totally rebuilt. Lina walked with Yarina and Ivan to the edge of the village. "Write to us as often as you can, won't you," Lina reminded her brother.

The cart with the horses was getting closer and closer. The men sat, holding onto all their belongings. The cart moved with extreme lightness because of the powerful strength of the horses as they covered the tracks, raising clouds of dust.

Women stood together watching the cart disappear into the distance, wiping the tears with their aprons. Not one of them started to walk back until the cart had completely disappeared from the horizon.

"Let's go, Mum," said Lina. "I know he is going to return. We are together, and I'm going to look after you.

"Yes, it's just me and you now." Yarina smiled.

Months passed, and they had no news from Ivan. Finally, a letter arrived. Lina opened the letter and read scribbly messy writing in ink. He told them how hard his work was. Every morning without fail at five o'clock the hard labour began, and they worked till eight in the evening. They had only occasional time off.

He told them how men of all ages—young and old—were getting on sharing their expertise. Everything had to be done manually because there was no machinery. They had only horses and carts to transport railway ties, tracks, and stones. They slept in hovels. Most nights they sat by the fire singing and talking to each other, telling jokes. Some drank, and some got so drunk they initiated fights. For others it was amusing to see men hitting each other till their noses bled. Anyone who got up and stood between them to stop the fight had a good chance of receiving a black eye, so eventually no one bothered to pull the combatants apart. Some cheered the fighters as if it was an entertainment.

After Lina reading a few pages aloud, she and her mother both breathed sighs of relief. They were just glad that Ivan was okay. Ivan informed them that he had been put in charge of taking attendance—putting ticks next to each man's name on the work roster every day. He wrote, "When I return, we will be able to do a lot of things—grow crops, collect the harvest." Lina proudly read the letter to her mum. It was evident by the tone of her voice how proud she was of her brother. "Mum, he is just like Dad!" she said.

Both women worked hard. Now they had to do even the jobs that Ivan used to do. Their workload increased and became much harder. Despite all of this, they were still managing.

They were not the only women in the village who had to work so hard. Many men and some women had left to help rebuild the country's infrastructure. Some had volunteered to work far from home. Most women in the village were widows or lived with disabled men. The war had brought so many tragedies to their families. And now the ruined infrastructure had to be rebuilt, taking more people far from their homes. Many volunteered because the government paid for the labour. The damage had to be repaired as far as the Far East. Many had said goodbye and had left to earn money so they could once again start farming when they returned home.

Life became even more difficult for Lina and her mother when the roof began to leak. The weather had begun to deteriorate. It went from bad to worse, and they were left to cope with dampness and cold. Burning wood in the stove to keep them warm wasn't enough. Someone had to repair the roof. Lina climbed onto the roof and managed to cover the damaged area with a thin piece of metal. Temporarily it would keep the inside of the house dry.

The days, weeks, and months went by. There was no news from Peter. Lina waited for his letters. She thought about him day and night, and their final words and promises to each other. A day didn't go by when she didn't think about him. Many times she wondered what his life was like now. She wondered how he was and where he was, and what he was doing. She continued to hope that one day he would find her. Deep in her heart, she knew that she loved him, and he loved her too.

She felt as if her heart belonged only to him, and no any other man would make her feel the same way. She remembered his words back in Germany: "Wait for me. I will find you!" Many times she thought about the moments they had shared in Germany.

She loved him even more now that they had spent so much time apart. They had been separated by the cruel ruling of the soviet government post-war agenda. Her mind was wandering, and she thought perhaps he had been taken straight from the army to do the same type of work her brother was doing. The country needed strong men everywhere.

There were no letters for months, but she was determined to wait. She had made her mind up. That was how sure she was about her love and dedication for the man she knew she needed, for the man with whom she wanted to have a family. He was the one for her; deep inside she knew it. No situation or person, even in their separation, could change her mind. She was prepared to wait, living with her mum as they helped each other.

They carried out so many duties, even those of the men who were no longer there with them. Their workloads were heavy. Somehow, they had the strength to go on, to handle everything. Both women were so tired they could just about make it to their beds each night. As soon as their heads touched their pillows, they were asleep. Post-war living was all about recovery, rebuilding infrastructure, and getting back to a more normal way of life. Many villagers helped each other with the little they had as they did their best to cope with life's ups and downs. Epidemics of infectious diseases took lives; the fact that there were no medical professionals in the area made it worse. The urge to survive was their encouragement as they strove for better things to come, for a better future, when their loved ones would return and give them help and peace.

Winter had arrived. It was as cold as ever in Babrichi. Drifts of snow almost covered the little huts of Babrichi. Only the tops of the roofs were visible. By twilight, the animals were snug in their sheds.

Smoke drifted up from the chimneys of the little cottages, which were buried in the snowdrifts. On the river, next to a black hole in the ice, stood women with rags wrapped thickly around their feet. An old man walked past them with a rake over his shoulder. In the far distance, the woods were completely covered by the thick snow, which looked like a fluffy cotton blanket. On the abandoned farm ruins, concealed by the snow, was a herd of skinny horses accompanied by a few scrawny sheep. Groups of sparrows tightly squeezed together under the cover of roofs. White laundry flapped in the winter's dry, cold air. Icy tracks left by the carts were everywhere. A woman walked along carrying a bundle of twigs. She wore a scarf wrapped across her chest. A blizzard had engulfed the village with blowing snow.

In some windows, little sparkles of light were visible. From time to time on the river, sleds whizzed along, the loud voices of the people riding on them vanishing in the distance. In Lina and Yarina's cottage, textile work was underway. All the women were sitting on benches, each one at a spinning wheel making thread, which they gathered onto spindles. Yarina was an experienced weaver, and she was spinning her own wheel so fast, the others found it impossible to keep pace. Lina enjoyed these gatherings in their warm cosy cottage with the mended roof. All the visitors were keeping their spirits up by sharing stories, singing, and laughing together. A few kids were lying next to the stove listening to the interesting stories shared in the cottage. In the wintertime, spinning thread and weaving fabric were common and necessary crafts. Everything was handmade, and the skills of craftsmanship were passed from generation to generation.

Motrya walked into the cottage with her young son. He looked so cold, his skin had turned to purple blue. Lifting her arms she said, "So warm in here! Can we come in? It's just me and my boy." She looked around and said, "So what about your men? Did you hear anything at all from them? Are they coming home for the winter?"

"No, nothing, not a word!" said one of the women. "We can expect them any day now."

"On my way to here tonight, the blizzard was so heavy I couldn't see anything in front me," said Motrya. "Your gate is so completely covered, we found it almost impossible to get through. In an hour, it won't be possible to get in or out at all! It's freezing out there!" She shivered and sat down. "I thought to myself that we must get to you lot rather than sitting on our own in an empty house. Bad thoughts have been coming to my head lately!" She shivered again.

"So you don't like praying then!" Yarina said. "Always complaining—ah and oh—and never even thinking of spending some time to get things off your chest! For a gossip, everybody finds the time. Change the way you live, and the bad thoughts won't come to your head, Motrya. Every time I hear you complaining, what about me or what about her? Life is not just to cross a field." A few women nodded.

Lina was listening to the women. Her imagination was always captured whenever she listened to their stories, told to the rhythm of the spinning wheels. How wonderful it felt to be with all these people in the snug cottage. She loved winter evenings. Lina always prepared hot food before their gatherings—usually a freshly cooked modest soup.

Lina got up heading into the kitchen. Yarina had made everybody's favourite, borsch. Hot food was just what was needed. As she took the big pot off the stove, the aroma of beetroot wafted around the cottage. It was enough to raise everybody's appetite. "Okay. Let us all have dinner now!" Yarina invited everyone to the table.

The noise of the wooden spoons scooping the hot borsch from the bowls was the only noise in the house. After dinner, everybody decided to stay in the house for the night because of the blizzard, which had gone from bad to worse, making it impossible to walk in the heavy drifts.

The smoke coming from the stove had covered all the walls in black grime. In the winter months, there was a great deal of dust in the cottages from the fires and the weaving waste. Winter that year had been unusually cold, and all the villagers were suffering. Some had run out of firewood because it was impossible to store enough wood to last all the winter season. A group of villagers had go out to the woods to get more wood.

People tried to keep warm by sleeping close to the stove. Nobody had double glass in their windows, so the glass was always covered with thin ice. Wind often penetrated the houses. This was a typical winter for every villager in post-war Ukraine.

That night, the women settled down to sleep, some near the warm stove, others on the comfortable benches. Inside the house was silent except for occasional snoring. That was all that disturbed the silence of the night.

Lina couldn't fall asleep as she kept thinking about her daily tasks and how much better her life was now. She was thinking about her brother. *He should return home soon*, she thought, as the work usually stopped in the winter months.

Not a day passed by that she didn't think about the man in her life. He was on her mind more often now than ever before. She could remember the times when they were together. With such thoughts, her eyelids became heavier with every second until she slowly drifted off to sleep.

PART 8

TOGETHER AGAIN

CHAPTER 22

Spring had finally arrived. Lina woke early, thinking about the jobs for the day. Together they were doing well keeping the farm stock and looking after the crops. Much work needed to be done in the house. She went outside. The morning sunrise and the fresh morning smells of the meadows were the best part of her day. Back inside, she washed her face, put on her clean clothes, and started preparing breakfast for her and her dear mum. The hot buns, eggs, and milk were on the table in no time.

Yarina slowly walked into the kitchen with a surprised look on her face. "You've been busy, child!" she proclaimed

"Didn't want you to start the day on an empty stomach." Lina smiled with great enthusiasm. Her smile was the one thing that no one could never take away from her. "It's bad for you, especially at your age, to load yourself with so much work," she pointed out.

"I don't want to put anything in my mouth—not a crumb," Yarina said. "I have no appetite, child." Her facial expression changed.

Lina sensed something was troubling her. She knew her mum very well. Her mum wouldn't say anything when she was in pain. "But, mum, you must eat," she insisted.

"I know. But it all gets stuck in my throat." With her hand, she pointed to her throat, indicating where the problem was.

"I must take you to see the doctor." Lina was serious. No one would doubt the look on her face. When she made up her mind, no one could ever change her direction

"What's the doctor is going to do?" Yarina was clueless. She had never paid attention to herself; she was always busy thinking about her children.

"He'll give you a check-up, a thorough check-up, Mum." Lina raised her voice slightly so Yarina could hear. Her mum's hearing was deteriorating, and often she would ask Lina to repeat what she'd said. Lina had become used to this, and to speaking a bit louder.

"No time for doctors, my child," Yarina said. "So much work must be done!" she proclaimed as if almost begging for Lina's guidance. Then she picked up a bun and started to chew with the few teeth she had left. Lina had been concerned about her mum lately. She noticed that her mum's appetite had decreased, and she had been trying to cook for her so they could eat together.

"By the way," Lina said, "don't worry about doing too much on the farm. I will help you when I return from work." Lina hurried away from the house, holding onto her bag.

Yarina said nothing in reply; she just smiled.

The day seemed to be off to a good start, and Lina felt as if she was almost flying with unusual lightness and a good feeling inside.

When the house was empty again, Yarina got up and started to get on with what she had to do. Suddenly the door flew open, and the post woman was standing with a letter in her hands. She was smiling, and there was excitement showing all over her face. "It's for you, Yarina!"

Yarina quickly took the envelope from the woman's hand; in fact, she almost snatched it. She was thinking about Ivan, her son. In desperation, she almost ripped the envelop open without even noticing it wasn't addressed to her. It was an official letter. As soon as the post woman left, Yarina took the letter straight to the head of village farming, Vasiliy. She was semiliterate and couldn't read very well. She thought that it was the letter of importance. She knocked on the door with her bony fist. The voice on the other side of the door said, "Coming!"

"Good morning, Vasiliy! I came to ask you if you could read this letter for me. It looks like it's an official letter. Could be some important news."

Vasiliy quickly opened the envelope and started reading the letter. He lifted his eyes and said with the smile, "Yarina! Dance! A young man called Peter is coming to ask your daughter to marry him! It looks like we all have to prepare for the wedding!" He was laughing now.

That was how everybody in the village found out about Peter.

"O, dear Lord!" Yarina put her hands together in disbelief. "This is so unexpected! She didn't even tell me much about him. She said she had met him, but she didn't tell me much more."

"I don't know," said Vasiliy. "The young fella isn't from here."

"We don't know his family," said Yarina. "We know nothing about him."

Well, you will have to meet him and see what for yourself," said Vasiliy. "You're a woman of a certain age. You should know the right man for your daughter who will take good care of her."

"Oh, Vasiliy, Vasiliy. I have only two children! If she goes, who am I left with? I can't run the farm all by myself!"

"She won't want to leave you," he said. "I know Lina. She is very close to you. Both of you have been coping so well with everything."

Yarina didn't say anything in reply. She just nodded with the signs of tiredness. "I'd better go. My house is full of work." With these words, she left, closing the door behind. All the way home she thought all sorts of thoughts. How could she let her only daughter go? Lina was old enough to get married if she wanted to. If Peter was the right man, Yarina thought, then Lina could get married. In her mum's eyes, Lina was just her child, her dearest daughter, whom she loved with all her heart, with all her soul. Nothing could ever change that.

CHAPTER 23

As she was every morning soon after Lina left for work, Yarina was busy doing her house chores. When Yarina walked outside, she saw a soldier on the horizon. He appeared to be walking in the direction of their house. She stood outside until he got closer. The soldier stopped and said with a smile. "Hello. Are you Yarina Uschuk?"

"Yes, I am Yarina Uschuk."

"I'm Peter." He walked closer to her. He took her hand and shook it firmly.

"Since you are here, come inside the house." She opened her front door and let him in. "Lina isn't here. She's out. Come inside and make yourself comfortable. Did you travel from far away?"

"Yeh, too far."

"You hungry? Want something to eat?"

Something to drink, *pani* (madame) Yarina, if you would."

Yarina was pleased by her first impression of the young man. He was not just good looking, he was friendly and well mannered.

"Where is she?" he asked.

"She left for work not long ago. Anyway, what are we standing here for? Take a seat. I'm going to get the food out! You must be hungry from doing all that travelling. I will make something to eat for you. She hurried into the kitchen."

"Thank you, pani." Peter took off his hat.

Yarina started telling Peter about his brother when he visited earlier on in the year and how she had received the letter from Peter and had thought it was about her son.

To Peter, Yarina seemed to be a nice and caring woman. He saw a lot of Lina in her. They enjoyed having a chat. The hot food sent delicious smells through the cottage, making Peter's stomach rumble. It felt so surreal. He had been thinking for months and months about going to find his sweetheart. The journey on the train had taken days. There had not been much food, and it had been cold, but he'd been determined to get to her. Love had driven him. He was persistent in his desire to make Lina his wife. He knew she was the love of his life, the girl of his dreams. He wanted to make his life with her.

When he had said goodbye to her, it had not been by his choice, and now he was in her house. He couldn't wait to see her, but he decided to be patient. It would be more sensible to help around the house. He noticed that the frame around one of the windows needed to be mended, some of the floorboards were loose, and the door needed to be straightened. Peter was a good carpenter, and he knew exactly how to do all these jobs. He wanted to help anyway he could. "Thank you, panni, for the meal. I can't even remember when I last ate home-made food. I see that lots of work needs to be done around the house. Now what can I help you with?"

"Nothing," said Yarina. "You must be tired from travelling so far. Go and have a rest."

"No! I'm not tired! I slept on the train for days and days. I'm a skilled carpenter. Let me help you," he insisted.

"Yeh," said Yarina. "With no man in the house, it's hard to fix everything. There's so much work inside and outside the house." The more Yarina talked to him, the more she liked him. There was something about him. Perhaps it was his determination and caring character she sensed.

So Peter got busy doing the chores, and Yarina went to work in the field getting rid of the weeds that were threatening the growing vegetables. They relied on their farm produce; it was their only source of food. Working hard enabled them to feed themselves, and they could exchange some of their produce for grains, flour, oil, and coal. The hard work had paid off. The weather had been mild. Many were predicting a good harvest. For the villagers in post-war Ukraine, a successful harvest was so important—a life-and-death situation. If they didn't work, they didn't eat. It was as simple as that. They worked on their own farms and on the government farms. Everyone had to work hard to feed their families and contribute to the government.

As soon as Lina finished her workday, she walked back home with the girls who lived nearby. She was very happy as she talked with her friends, and they laughed and joked together. As soon as they said their goodbyes, she walked towards her house. When she walked inside, she sensed that something seemed different.

Her mum came out from the kitchen and said with the smile, "We have a visitor today!"

"Who is it?" Lina asked. She was quire curios.

Suddenly, Peter came out from the adjoining room. Lina made an exciting noise and ran to him, hugging him. "I'm so happy to see you!" she exclaimed happily. She bombarded him with millions of questions and couldn't believe that he was there with her after such a long time of waiting and worrying. She could hold him now and talk to him. Lina had missed him so much. Her heart was filled with love and affection towards this man. He looked so handsome and fit after his life the army. Lina told Peter about her life after he had been taken into the army. She told him about her long journey home. She told him about her risky journey from Germany and how glad she was to be home with her mum.

"So what happened when you left me in Germany? Tell me!" she asked with curiosity and excitement. She couldn't stop smiling.

"Oh, they transported us on the train to the Ukraine. The train didn't have any roof. All the wagons were open. So we were freezing, as there was nowhere to hide from the terrible wind. In the night it was horrible—cold and wet. As we entered the Ukraine, the train passed through a forest. These people who lived in the forest were hungry and wild. They threw ropes with loops tied in them to catch people from the train. It horrified me. We tried to avoid the ropes, but there was not much space. One loop got hold of my jacket, which was the only thing that protected me from the wind. I had no choice but to let the jacket go. After all, they would have pulled me off the train!"

"Oh, my God!" Lina's was terrified. "They could have killed you and eaten you! Those are wild cannibals who hide in the forest. They rob and kill!" Lina told Peter. "A few people have disappeared walking through the forest. No one has seen them since they went into the forest."

"Yes," said Peter. "I was terrified and glad at the same time that they got hold of my jacket and not me!" He smiled so he wouldn't worry her. Lina smiled back. She loved his sense of humour.

"When we reached our destination," said Peter, "I was given an army uniform. I was able to have a wash, and I changed into my new clothes. Fresh as a cucumber I felt. Then the training began. In the morning we started our new regime. The training was tough, and we learned lots of new skills. I missed you all this time. You were on my mind constantly." He lifted her chin, and their eyes met. They moved towards each other and shared a passionate, long-lasting kiss. "I want to be with you forever," Peter said with stern look.

"Me too," Lina replied softly.

Yarina walked into the house and unintentionally interrupted their tender moment. "I know you have lots to talk about, children, but we must have dinner, and after the dinner you will have all the time in the world to catch up."

"Okay, Mum. Let me help you."

"I'll bring some logs into the house and start a fire in the stove," Peter said while two women were busy in the kitchen.

"It seems you already know what to do without me telling you!" Lina said to Peter with a smile.

Dinner was great, and they talked about lots of different things. The two women were captivated by his stories. He was a very interesting person to have a conversation with, and he had so much to tell.

As soon as dinner was over, Peter and Lina helped Yarina to tidy up the table so they could carry on with the conversation. They talked until the early hours of the morning and couldn't believe how time quickly passed.

Peter took Lina's hands and said, "You look tired, my darling. Perhaps you need to go to bed now," he suggested.

She was tired but didn't want to admit it. Her excitement was much greater than her weariness. She couldn't keep up her eyes open. Her eyelids were getting heavier and heavier.

Peter was going to sleep in the living room on the stove bench. Yarina had allocated that warm place for him so he could have a good rest. "You must rest for today," Lina said. "I must be up soon. Actually I have only four hours to sleep. I must turn up at work on time. The punishments are severe if I don't."

She got up after kissing him gently on the lips and walked to her bedroom closing, the door behind.

Lina quickly fell asleep thinking how happy she was that Peter was finally there and he had come to see her as he had promised her back in Germany.

Peter couldn't get to sleep, thinking about Lina and how much he loved her and wanted to be with her. He loved everything about her—her lovely smile, her beautiful green eyes, her character, her bubbly personality, and her freshness. He wanted to be with her for the rest of his life, and he hoped she wanted to be with him. He decided to propose to her. He made his mind up to do it the next day. With this thought he fell asleep.

CHAPTER 24

Peter's and Lina's wedding ceremony was a small, and only close relatives were present. They celebrated their special day by sitting at the big table in the circle of their family. Lina's uncle played an accordion. Everyone danced and sang all night. They hadn't had a good time like that for a such a long time. The newlyweds were happy that the moment they had been waiting for had finally arrived. They were now husband and wife and committed to spending the rest of their lives together. Their future looked brighter because Peter now had a job. Vasiliy, the head of farming in the village, liked him and had given him a responsible position. He was put in charge of the barns where the crops were kept, and he had various duties to perform. They were happy for what they had. Peter couldn't take his eyes off his beautiful wife.

Lina's eyes sparkled, and she looked like a star in her white dress and beautifully hair plaited with carefully arranged fresh flowers. She looked amazing.

Lina looked at Peter lovingly. She couldn't believe he was now her husband. She loved everything about this man. He was nice, and he was magnificent. She was very happy to be his wife.

Ahead of them was the long road of life. Neither Peter nor Lina knew how life would treat them, but they had each other. They believed in one another. And they believed in building their life together and starting a family.

The horrid, terrible, and tragic times of the past, brought by the war, had left many psychological and physical scars. Neither of them could ever erase the memories of the most devastating war. Despite the troubled past, however, they had found love. They thanked God daily for delivering them alive from the dreadful nightmare. Their love would grow and flourish. They would be blessed from that moment on as husband and a wife for their loyalty and hard work.

ABOUT THE AUTHOR

My name is Olga Nunn. I am originally from Ukraine, a country that used to be under the hard ruling of the Union of Soviet Socialist Republics (USSR). My school years left me with the skills I needed to write. I love writing stories and short articles based on my research and observations. I have managed to publish some of my articles. My passion for writing has grown over the years, as I am good at observing nature changes. I am very keen to convey stories that are captivating to my readers.

ABOUT THE BOOK

This story is about my Ukrainian grandparents, who met in Germany during World War Two. Their story is about a love that began during the most devastating and eliminative war in human history. They were both taken from their families and put into slavery in Germany. A vast number of young people were victims of this political tactic against their will. Despite this political agenda, Lina and Peter found each other and found love that would turn separation to togetherness and create a new life in post-war Ukraine.

Made in the USA
Monee, IL
04 November 2026

67375610R00105